LONE STAR
SUBURBS

LONE STAR
SUBURBS

LIFE ON THE TEXAS
METROPOLITAN FRONTIER

EDITED BY PAUL J. P. SANDUL
AND M. SCOTT SOSEBEE

UNIVERSITY OF OKLAHOMA PRESS : NORMAN

This book is published with the generous assistance of the McCasland Foundation, Duncan, Oklahoma.

Library of Congress Cataloging-in-Publication Data

Names: Sandul, Paul J. P., editor. | Sosebee, M. Scott, 1961– editor.
Title: Lone star suburbs : life on Texas's metropolitan frontier / edited by Paul J. P. Sandul and M. Scott Sosebee.
Description: First Edition. | Norman, OK : University of Oklahoma Press, [2019] | Includes index. | Summary: "A collection of essays exploring suburban development in Texas. Essays consider transportation infrastructure, urban planning, and professional sports as they relate to the suburban ideal; the experiences of African Americans, Asian Americans, and Latinos in Texas metropolitan areas; and the environmental consequences of suburbanization in the state"— Provided by publisher.
Identifiers: LCCN 2019019905 | ISBN 978-0-8061-6447-2 (paperback)
Subjects: LCSH: Metropolitan areas—Texas. | Cities and towns—Growth. | City planning—Environmental aspects—Texas.
Classification: LCC HT334.U5 L66 2019 | DDC 307.76/4—dc23
LC record available at https://lccn.loc.gov/2019019905

For our friend Ted Lawe,
truly a gentleman and a scholar
Rest in Peace, Friend

CONTENTS

ACKNOWLEDGMENTS

One of the biggest benefits of academic history conferences—or any gathering of professionals within the discipline—is that in our conversations, questions, and answers at sessions, or even just sitting in the lobby of a hotel discussing what it is that we do, we often discover the genesis of a book or study. Although *Lone Star Suburbs* did not begin at a conference, it had a somewhat similar creation. We are both football fans, and we also have the misfortune of pain that comes with having the Dallas Cowboys as our favorite team. So, we sat one day watching the Cowboys and, probably because Dallas was losing, also discussing history. Specifically, we were discussing Texas history, the evolution of the state in the twentieth century, and how it had undergone such great change in the latter half of that century. We observed that the tale of twentieth-century Texas was one of a transformation of the state from a predominantly rural place to one characterized by great urban spaces, a revolution that many historians and geographers have noted. But what also dawned on us was that Texas's urban regions were of a certain nature, that they almost all fell into a certain urban category: Texas was a suburban state. We became excited—that may have somewhat been the result of the adult beverages we had consumed—because we also realized that this gave us a niche to study. We had a project!

We discussed several ways to approach such a study; perhaps a monograph was best, but since very few if any works had focused on Texas's suburban processes we were sure that was the "conversation starter" we desired. Thus we settled on an anthology, and then we faced the task of finding the proper authors to write the essays. It took us nearly a year, but we think we gathered a roster of the perfect people not only to examine the various suburban processes in Texas but also to spur the questions that will influence other scholars to explore this vital topic. Texas is a suburban state, and though the cultural image of Texas usually involves cowboy hats, oil derricks, football, and loud, brash people, that is an overblown stereotypical image. If one wanted a true "Texas image," one would place its people in single-family dwellings, in planned developments, where one would be surrounded by shopping centers, chain restaurants, and an almost homogenous culture. You want to find a Texan? Odds are, you would find him or her in a suburb.

Our hope for this volume is twofold: we want to explore the process of how Texas became a suburban state; but we also want it to lead to other studies, other scholarly investigations that will deepen our understanding of the Lone Star State in the twentieth century, a region that has undergone a fantastic transformation into a land of suburbs.

I am thankful to *all* my family and friends. Each of you knows who you are, so sorry if I don't mention you by name (sorry, Book Club buddies), but I would especially like to thank my wife and son, Tosha and my precious Max, as well as my mom (Diana), dad (Duane), and brother (Glenn). Thanks also to my colleagues in the history department at Stephen F. Austin State University and around the SFA campus (and beyond) who listened to me rant and rave about this project: Mark Barringer, Brent Beal, Heather Olson Beal, Perky Beisel, Lisa Bentley, Chris and Jamie Bouldin, Lauren Brewer, Philip Catton, Kyle Conlon, C. C. Conn, Dana Cooper, Troy Davis, Daaron and Shonda Dwyer, Ashley and Saville Harris, Bill Nieberding, Brook Poston, Amanda and John Pruit, Steve Taaffe, the Wenner family, and Carrie and Stephen Wright. Of course, the whole of the history department provided tremendous support, no doubt, including Sue Dockery, but I give special thanks to Andrew Lannen for reading over and editing my chapter.

This project was and is also very personal for me. I was born and raised in an American suburb. In many ways this project continues my broader investigation into what helped make me me. Suburbs have simply been my major academic focus, but I have now moved from a robust focus on the Golden State suburbs of my youth to the Lone Star State variety of my son's. Last, thanks to Scott Sosebee, my colleague, my dear friend. Nevertheless, whatever story he may tell you, this was all his idea.

PAUL J. P. SANDUL

I am decidedly not an urban, or a suburban, historian; rather; I am a Texas and southern historian. But my colleague, most of all friend, Paul J. P. Sandul, keeps trying to make me one. Despite such an admission, I have to thank him for pushing, prodding, and guiding so much of this project because I have come to enjoy it, and it has been one of the most satisfying educations of my life. Thank you Paul, for making me a better historian. I also must thank my colleagues in the Stephen F. Austin State University department of history. Paul named you above, and without your support and encouragement this project would not have come to completion. I also extend great kudos to my friends and fellow historians Ty Cashion and Gene Preuss. Both of you were always there for me to bounce ideas off, to help me with the process, and to help me understand so much about the needed direction of Texas history. Thanks, guys, for all you have done. I have thanked them in other mediums, but I must always extend my gratitude to my mentors, Paul Carlson and the late Donald Walker. Without your guidance, friendship, and sometimes needed kicks in the behind, I would not have made it. I owe my professional existence to you. Of course, most of all, I acknowledge the person who motivates me most because she always supports me, never lets me get down, and is my biggest fan. Leslie, you are not only my life partner but my best friend.

M. SCOTT SOSEBEE

LONE STAR
SUBURBS

{ 1 }

TEXAS SUBURBIA RISING

PAUL J. P. SANDUL

Texas is a suburban state. It is not a rural state nor even really an urban one. More rightly Texas is a metropolitan state. In the 2010 federal census Texas had two of the largest five metropolises in the United States: the Dallas–Fort Worth–Arlington metropolis (ranking fourth with about 6.5 million people), and the Houston-Sugarland-Baytown metropolis (fifth with more than 6 million people). San Antonio–New Braunfels, with more than 2 million, ranked twenty-fifth, and Austin–Round Rock–San Marcos ranked thirty-fifth with just less than 2 million. According to the census bureau again, out of 366 nationally in 2010, Texas had twenty-five metropolitan statistical areas (MSAs)—areas surrounded by and including a central city or twin cities of more than 50,000. To put this in perspective, in 2010, 22,526,822 of the 25,145,561 total state population lived in Texas's twenty-five MSAs. So almost 90 percent of all Texans live in metropolitan regions that are primarily made up of both cities and suburbs. But, as we shall see, Texans mostly live in suburbs.

Seeing Texas as a metropolitan state provides us with a different way of thinking about Lone Star history. Take the old argument about whether Texas is a southern or western state—an argument that has long pervaded the state's historiography.[1] Does the past matter most? If so, which past? Do Confederates in the attic, Jim Crow, lynching, and the Klan up through the civil rights movements of the 1950s and 1960s forever mark Texas as southern? What about the Alamo? Do cowboys and Indians, cattle trails, farming, Texas Rangers, and outlaws forever mark Texas as western? What about the present? Does a preference for, say, sweet tea, hush puppies, and saying "yes, sir" and "no, sir" mean you are indeed southern? Beef! If it is what's for dinner are y'all western? Does location on a map even matter? Perhaps Texas was and is both. West Texas is western and always has been.

East Texas was southern and always will be. All of Texas is a mix, if not God's country. Perhaps it has and currently still does depend on your race or your gender, religion, or distinct points of view and life experiences.

Seeing Texas as a metropolitan state complicates the southern-western debate even if it does not necessarily erase it. But Texas has included an enormous suburban population, especially since World War II. Texas's suburbanism thus links it to a broader American suburban history and contemporary way of life. For all their real differences, the metropolises of Austin, Dallas, Houston, and San Antonio share historical roots and modern-day phenomena with Boston, New York, Chicago, San Francisco, and Los Angeles. This does not mean that Texas is or is not southern or western—or distinct. It just means that, as historian Robert A. Calvert once wrote, "the ambivalence concerning Texas as either Southern or Western has let the state fall between the historical cracks of some of the current national and regional scholarship."[2] Following a suburban history and reality means that Texas reflects national trends as much as parochial ones.

TEXAS HISTORIOGRAPHY, THE TEXAS MYTH, AND MISSING THE SUBURBAN

Texas historiography has yet to catch up with a suburbanizing metropolitan state. This is not at all surprising inasmuch as popular coverage about suburban Texas, particularly in the media, is recent.[3] Modern works on Texas's past also make clear that Texas has had its own historiographical trends that have dominated and shaped the state's history.[4] Though each work brings its own insight to the history of Texas historiography, they all make clear that John Wayne rules—be it *The Searchers* John Wayne, *The Alamo* John Wayne, or *The Comancheros* John Wayne. Texas history, in other words, has historically been limited in scope (nineteenth century) and dominated by cowboys, the frontier, Texas Rangers, and above all the Alamo and the Texas Revolution—a.k.a., the "Texas Myth." Political history has recently emerged and extended the chronological coverage of Texas history beyond the nineteenth century but not to the topics, phenomena, and, pivotally, people fully representative of the Lone Star State's past and by implication present and future.[5]

The narrow focus on the Texas Myth is actually a widely successful construct. In simplest terms, historians and others who had previously decided on such a mythic focus built the Texas Myth. The Texan past is not what happened per se but what these interpreters have said happened. This line of reasoning admittedly follows a view that representations of the past, especially history, are inherently narrative and imaginatively constructed. Moreover, history is not just about making a subject objectively known but also crafting certain representations that often convey both surface-level meanings and deeper myth-like connotations. This does not mean so-called historical truth is impossible to capture, just that history is constructed and open to subjective interpretation and at times misrepresentation.[6]

Past historians also often function similar to what some scholars have called "regimes of truth," "regimes of representation," or "experts in legitimation."[7] The idea here is not to introduce needless jargon or snootily footnote a French theorist but to underscore how such regimes or experts connect to systems of power which, in one French theorist's words, "produce and sustain it, and to effects of power which it induces and which extend it."[8] In short, these regimes or experts (such as historians) reflect a structure and relation of power that establishes so-called truth and disseminates it. Past Texas historians, in this view, have worked to construct truth. They have fashioned stories about Texas history, often infusing them with widely popular and pertinent cultural symbols and characters (discussed below).

Seeing history this way enables us to cast the Texas Myth as "dominant memory," which simply refers to conceptions of the past that acquire dominance in both public and academic representations.[9] I do not mean to single out and indict Texas historians, however. They certainly did not act alone—and I do not want to receive any of their hate mail. Historians and their writings are but one of what some have called many "theatres" or "vectors of memory."[10] Clio, they say, has many invisible hands.[11] History is surely not the sole domain of historians—and despite the lament of many historians they are likely not the most influential.[12] Representations of the past abound in television, film, photographs, novels, the internet, video games, newspapers, magazines, memoirs, oral histories, local history books and societies, blogs, dictionaries and encyclopedias, museums, art, architecture, pageantry, monuments, memorials, historic preservation, and commemorations.[13] Yet, as many of the relatively recent surveys concerning Texas historiography have made clear, dominant memory directs both Texans and others to think of Texas's past in ways that reify, legitimate, and disseminate the Texas Myth.

This repeated and faithful reconstruction of the Texas Myth throughout time and across various representations can be surprising for some. Why so dominant? Besides the stories being entertaining to their audiences (who doesn't like John Wayne?), notions of "collected memory" and "invented tradition" encourage us to see that earlier representations provided by past historians and others became the resources of others, whose works and representations then became the resources used by others, and so on.[14] The Texas Myth therefore does not so much result from Texans or others lacking creativity or consciously deciding to perpetuate it. Rather, it is because other ways of representing the past in and about Texas seem nearly impossible to do outside the ways it has previously been represented or remembered.

The Texas Myth as both dominant and collected memory certainly casts a long shadow. Scholars of memory have tied memory to a set of practices like commemoration and history writing precisely to underscore the invention of tradition, dominant memory, and personal and community identities.[15] Said differently, the Texas Myth helps to shape people's understanding and knowledge of the past—their historical consciousness—and a sense of self, what one developmental psychologist calls "narrative identity"—"the internalized and changing story of your life."[16] This evolving story includes the engagement, consumption, reproduction, and reassembling of representations of the past such as the Texas Myth.[17]

Another important function of history and memory (e.g., the Texas Myth) is that they have "a remarkable capacity to create a sense of unity or 'oneness' among people who would not otherwise see a meaningful sense of kinship."[18] At the level of the group, a dominant or popular memory can function to bind Texans "together by instilling in them a sense of common mission and destiny."[19] The goal is to create a shared likeness among differing, complicated individuals so that the group in which they live is (or at least appears) further united and stable—often labeled an "imagined community"—in other words, the invention of tradition as an essential basis for the creation and maintenance of groups.[20]

The Texas Myth, though an accomplishment, does also reflect neglect, particularly concerning the place of women and minorities (not to mention their possible embrace of the Texas Myth). It also overlooks urbanization and much of the twentieth century. But what is also clear after reading Texas historiography is that this common lament is no longer as true as it once was—at least in academia. An increasing number of high-quality works have appeared on women, minorities, cities, sport, popular culture, religion, politics, and the twentieth century, suggesting a "new Texas history" that is not content with myth, misrepresentation, and white men.[21] Texas historiography thus helps us to understand both why suburbia has largely been left out of the old Texas Myth and why suburban history is now poised to break out as part of a new Texas history that moves beyond the mythic. If the dominant Texas Myth still does influence Texas history (and it does), then the more recent new Texas histories are doing the same. The construction of history is not static. However powerful dominant or collected memory can be, with every reconstruction of the past there is a chance for change. A new Texas history, made by the collective word processers of contemporary Texas historians, certainly reflects and aids this ongoing reassembling and repackaging of Texas history to which suburbia can now join.

Texas historiography is not entirely devoid of suburbs, however. For instance, Texas's growing urban history is exciting, reflecting as it does a new Texas history that promises to grow into a more encompassing metropolitan history that places cities and suburbs within larger geographical, political, demographic, sociocultural, and economic trends. No understanding of cities can come without suburbs, and vice versa. Richard B. Wright's review of Texas urban history in *Discovering Texas History* makes clear that Texas has a long tradition of urban biography and county-level history. Wright also puts it bluntly: "Most all of Texas history since the end of World War II is also an urban history by definition," because most Texans have lived in cities since then (or, more correctly, most live in suburbs). More recent urban history in Texas has continued urban biography (and a recent suburban biography of Highland Park and River Oaks), especially about Houston, Dallas, and San Antonio, though recent studies on Austin, Brownsville, El Paso, Fort Worth, Galveston, Lubbock, and Waco have appeared. What is also impressive about these histories is a new Texas history approach that includes minorities, women, and even the environment.[22]

Despite the virtues of these new Texas histories, suburbia itself is not covered in detail. They are mainly political histories that seek to understand the rise of conservatism in Texas, then situate it in suburbs and the so-called suburban lifestyle of "Little Boxes on the Hillside." This is the case with Linda Scarbrough's *Road, River, and Ol' Boy Politics* (2005), Bill Bishop's *The Big Sort* (2008), and Sean P. Cunningham's *Cowboy Conservatism* (2010).[23] They surely paint a compelling picture of conservatism's rise within Texan (and other) suburbs by the 1970s. In these neighborhoods, (neo)conservatives began to rise, organize, self-segregate, and mobilize for alleged law and order and fiscal responsibility in the face of a tumbling economy, stagflation, civil rights violence, integration, Vietnam protests, and the counterculture's so-called glorification of drugs, sex, and rock n' roll. Neoliberal trends of deregulation, privatization, and the marketization of nearly all aspects of life gained a footing, first in conservative think tanks such as the Heritage Foundation, then in the administrations of both Jimmy Carter and Ronald Reagan. Watergate also proved what most knew—that the government lies. Suburban evangelicals mobilized too, particularly in light of the Stonewall riot in 1969 and the emergence of LGBTQ rights, the Equal Rights Amendment, second-wave feminism, and *Roe v. Wade*. Indeed, suburbia came to represent a white, conservative "silent majority" by the end of the decade as Texas, in 1978, also elected its first Republican governor since Reconstruction.

ANTI-SUBURBANISM, THE RURAL IDEAL, AND MISSING THE SUBURBAN

Despite the presence of suburbs in the aforementioned studies, suburbia as a whole remains underanalyzed in Texas. The prevalence of the Texas Myth as both dominant and collected memory is one reason for this. Two other reasons seem culpable as well: a larger American disgust with suburbia and all it supposedly represents, and the power of the rural ideal in a state that is decidedly not rural anymore. Both have coupled with the Texas Myth to help camouflage and constrain the study of Texas suburbia.

The definition of "suburb" is more complicated than it seems at first glance; suburbs are simply too multidimensional and varied. Recent historical literature concerning suburbia has documented industrial suburbs, racial and ethnic suburbs, multifamily housing, and more, moving us beyond the white middle-class residential archetype. Yet, when pressed, suburbs are simply places. They are places somewhere near a city or cities (whether directly tied to the city core or not) with modest residential density and a plethora of services and leisure facilities; hence, whether true or not, the long-standing trope of using a Starbuck's or Olive Garden as both symbols and symptoms of suburbia. Yet we need to tread lightly and not be too rigid in the definition of the term, as if suburbs, in reality, have been constantly one thing. Suburban form and function and suburbanization as a process are contingent on historical phenomena and context and are dependent on the people living there. They both can and do evolve.

A major difficulty with trying to define the suburb is the emphasis on physical location and place, which risks downplaying the importance of imagery and imagination, of what some have called "storied space." A suburb is as much a cultural symbol and intellectual creation, even lived space, as a geographical or material place. Suburbia's critics themselves have made this abundantly clear as they have vociferously bemoaned American suburbs for flattening out individuality and eroding civic participation.[24] Critics like William H. Whyte and his *Organization Man* (1956), Betty Friedan's gloomy "problem that has no name" (1963), and James Howard Kunstler's proclamation (1993) that "we created a landscape of scary places, and we became a nation of scary people," have made suburbs what suburban scholar John Archer has called "The Place We Love to Hate."[25] Suburban critics popularized the belief that suburbs darken our very souls just by living in them. In this scenario suburbs are a vulgar cultural product of mass society in a consumer-capitalist world. They are often scapegoats and wrapped in stereotype and mythology.

Certainly popular cultural depictions on television like *Leave it to Beaver* (1957–63) and *The Brady Bunch* (1969–74) have broadcast a more positive image of suburbia, even when they simultaneously reinforced the white middle-class residential trope, complete with gender stereotypes (*Father Knows Best*, 1954–60, after all). Collectively they presented a narrative of the American family in suburbs who represented proper family values and relationships, appropriate gender behaviors, and a tireless, patriotic devotion to strong moral character.[26] Yet, joining suburbia's critics, some movies like *Mr. Blandings Builds His Dream House* (1948) and *Rebel without a Cause* (1955) satirized life in the suburbs (just think of Dean's character crying, "You're tearing me apart!"). In the 1970s, as the United States became a suburban majority for the first time, fictional psychopathic serial killer Michael Myers, born and raised in suburbia, returned from the mental hospital to kill the babysitter in John Carpenter's 1978 release of *Halloween*.

More movies and popular cultural representations exist, of course, from Bob Hope's *Bachelor in Paradise* (1961) (where he plays an international travel writer investigating exotic American suburbia and finding love) to Jane Fonda and others in the 1962 release of *The Chapman Report* (whereby Fonda's lack of libido apparently kills her suburban husband) and, in 1979, *Over the Edge* (depicting sex-crazed middle-school drug dealers in suburbs). We can add, among a plethora of others, *Stepford Wives*

(1975), *Poltergeist* (1982), *Nightmare on Elm Street* (1984), *Breakfast Club* (1985), *The 'Burbs* (1989), *The Truman Show* (1998), *Pleasantville* (1998), and *Disturbia* (2007).

On the one hand, suburbs can surely stand for home, family, and community. On the other, they can also signify conformity, uniformity, consumerism, alienation, impedance, racism, sexism, homophobia, and elitism. They are also maladaptive; they are not equal to the task of raising well-adjusted boys and girls to become proper American men and women (think of Kelly and Bud Bundy in *Married with Children*, 1987–97). They are not good for marriages (*American Beauty*, 1999), safety (*Last House on the Left*, 1972), or even masculinity (Dustin Hoffman's Ben Braddock in *The Graduate*, 1967). Even Ice Cube thought "the suburbs make the hood look good" in *Next Friday* (2000). Ultimately, American anxiety over suburbia abounds and in all likelihood plays into Texan anxieties as well, particularly since it runs counter to a Texan ethos of rugged individualism and distinctive (but noble) peculiarity. Because of this, Texans may very well be hesitant to identify the suburban in their state precisely because they do not like it, do not want to admit it, and, in due course, reject it and all it has been represented to stand for.[27]

What Texans do embrace, however, is proverbially complex. Not surprisingly, given what suburbs have come to represent popularly, many Texans revel in alternate representations and ways of thinking about both the past and the present—of where they live. Of importance here is the power of the rural. With nearly 90 percent of the Texan population living in metropolitan regions, it might seem counterintuitive to talk about Texas and the rural now. Nevertheless, many Texans embrace rurality, or at least a Texas version thereof, as a cornerstone of state identity and propagate that image to both themselves and the world. So although Texas is overrun not with cowboys or J. R. Ewing and oilmen but with so-called soccer moms, hipsters, and NASCAR dads who like to shop at super regional malls, it endures in both dominant memory and imagination as a very rural state. Here the rural ideal holds powerful sway, and in the Texan version primarily through the iconic image of the cowboy and the republic for which he stands. And that republic, mind you, is not always the American one but the blessed Texas Republic of lore.

At its core the rural ideal assumes and maintains a sense of superiority. The championing and privileging of rural environments is old, however, likely a tale we can spin as far back as the ancient Roman poet Virgil or the Hebrew religious legend of a Garden of Eden. Some evolutionary scientists and psychologists even argue that human beings are predisposed to favor rural environments because of our evolutionary heritage and longtime growing up as a species in the wilderness.[28] Yet a more modern celebration of the rural emerged with the rise of romanticism, first in western Europe in the mid-eighteenth century, then in the United States by the early nineteenth century. Romantics such as Ralph Waldo Emerson, Henry David Thoreau, and even Thomas Jefferson lionized a sacred "rural virtue" while decrying the effects of industrialization, urbanization, and centralization. Though cities often received praise from local urban boosters, romantics contributed to an early and growing hostility to the city. Cities, in their view, were a menacing and, akin to anti-suburban rhetoric today, maladaptive environments—bad for morals, families, health, and prosperity.[29]

According to romantics, a home in the countryside provided families a safe and simple gathering place in an environment of natural, healthful, and homogenous surroundings. Picturesque rural cottages and villas like those of Alexander Jackson Davis or Frederick Law Olmstead became commodities eagerly consumed by middle-class Americans, giving birth to not just the rural ideal but also, as Jake McAdams reminds us in his chapter in this volume, the suburban ideal. This rural antidote to a growing urban disease celebrated an agrarian past, linked democratic citizenship with rural life, and represented a place for experiencing what it meant to be an American.

Despite demographic realities, many Texans embrace a rural image and "rural virtue." The rural ideal casts those living in such places as "real Americans," the backbone of both the American and Texan republics (or, as Waylon Jennings sang, "You just can't live in Texas unless you got a lot of soul"). This is the familiar agrarian orthodoxy of farmers as society's morally superior heroes, but adapted and applied to all rural people (real or imagined, farmers or not).[30] Rural Americans, of course, are often cast as country bumpkins, a backward electorate who vote against their own interests and ensure America's unremitting misery, embarrassment, and stagnation. They talk and dress funny too. Yet many Texans appropriate the rural as a decent, moral, conservative counter to urban, liberal America. They invoke the rural image as a way to mark themselves as different and as better than people from other states. Texans see their rural but urban state as distinctly *not* California or New York. They are better, more American, and superior, or so many of them say. Here, the urban represents what, speaking critically of cities, Thomas Jefferson condemned "as pestilential to the morals, the health, and the liberties of man."[31]

Texas has its own brand of the rural, of course, molded and shaped by Texans and for Texans as much as it has been inherited or replicated by dominant and collected memories. Specifically, Texas is the home of the mythic cowboy, a central figure in the Texas Myth and the American West tradition. As historian Paul Carlson and others have made clear, rurality in Texas is a mix of the romantic rural ideal (the noble, morally superior agrarian), Frederick Jackson Turner's Frontier Thesis (whereby whites overcame "savages" and sagebrush to ensure the continuance of democracy), and the cowboy (armed and dangerous). The Texas cowboy became a debonair hero who tamed a savage West and lived off Mother Nature, be it her plentiful bounty or, even better, dry and barren nothingness and thus requiring both brains and brawn to survive. The cowboy is a true man's man.[32]

The heyday of the cowboy in popular culture came in the 1950s and 1960s amid both the cold war and classical phase of civil rights. As Judith and Andrew Klienfeld, Andrew C. Baker, and other scholars remind us, the cowboy became a cold war icon depicting American exceptionalism and triumph over nature, race, and the Soviet Union. The United States embraced an image of a rugged yet moral individual who sought private gain on the one hand but built an entire democratic and capitalist nation on the other. Quick to thank the Lord for both his supper and horse, to tip his hat to "the ladies," or right the wrongs of local ruffians, the American (or in this case Texan) cowboy was a powerful alternative message in a nation that would not allow its nonwhite citizens to enjoy the full blessings of liberty. Segregation, discrimination, and more were America's proverbial Achilles heel in a cold war rhetorical arms race about global moral superiority. The cowboy, though, represented American righteousness, scruples, and success.[33] For Texans, in particular, as historian Randolph Campbell has argued, the cowboy became a way to escape the "burden of Southern history." The cowboy way evoked a vision of a pilgrim's noble errand into the wilderness rather than a past drenched in African American blood, from the stain of slavery to the evils of Jim Crow. This also is why Texans turned toward a more overt celebration of the Texas Revolution and downplayed (though never eliminated) the Confederate past.[34]

Type "cowboy" in a Google image search and what stares back at you is a white man dressed in a cowboy hat and boots, usually with a vest, gun, and riding a horse (or a Dallas Cowboy). Make no mistake: the cowboy is white and macho. Despite the fact that we know most cowboys in Texas and the West were not historically white but rather African American and Hispanic, the narrative image of the cowboy is unquestionably white. This is no mistake, either. This was a purposeful invention of tradition whereby the cowboy "reaffirmed masculinity, conquest, and white privilege during a time of Cold War uncertainty and racial tension."[35] The cowboy represented a counter to a supposed feminization (and thus corruption) of modern society. The cowboy thus emerged as an archetype of American masculinity: hardworking and in tune with nature but still intelligent and devoted to family, God, and respect (which he demands or he will shoot you).

Although the heyday of the cowboy in popular culture came in the 1950s and 1960s, even as President Ronald Reagan embraced the cowboy as emblematic of conservative values and rurality as synonymous with whiteness, the broader appeal of the cowboy had subsided by the 1980s.[36] Still, the cowboy rides high in Texas. The reasons for this are many. The simple love of good stories can get the best of any of us, no doubt. The power and influence of a dominant Texas Myth is another reason, as is anti-suburban rhetoric and disgust. The cowboy according to Texas and western mythos represents rurality, moral superiority, conservatism, faith, masculinity, and white worldviews. As Texas grows more suburban and racially diverse (and Islamic and "other" faiths too), then perhaps no surprise should come from the continued popularity of the cowboy, particularly among white Christian Texans. The rural and Texas cowboy myths allow Texans to continue to deny their suburban reality, maintaining that they are still rural in spirit.

I have highlighted the power and influence of the Texas Myth, anti-suburbanism, and the rural ideal. It is my assertion that such ways of thinking have resulted in Texans minimizing the suburban side of their history, whether consciously or unconsciously. Dominant and collected memory in the form of a Texas Myth has favored nineteenth-century historical remembrances, narrowing on cowboys, great white men, the frontier, and the Texas Revolution. Even when extended into the twentieth century, Texas historiography failed to account for the suburban. What makes this frustrating is that, in fact, Texas is a suburban state, as the census data clearly show.

SUBURBAN CENSUS DATA IN TEXAS

Let the demographic facts be submitted to a candid world. But first a caveat. The census bureau does not rigidly define suburbs. This makes our task a bit harder, though not impossible. According to bureau definitions, suburbs are unincorporated and incorporated communities of moderate density typically located outside a central city or cities. Suburbs reside within a metropolitan area and are under an urban center's sociocultural and political-economic orbit. This means the bureau designates the largest or best-known cities in most metropolitan areas as a "central city," including, for example, the cities of Dallas and Fort Worth, but also Irving and Arlington (and not treating any one as a suburb). Consider just Irving in the Dallas–Fort Worth–Arlington metropolitan area, which is labeled as a central city despite the fact that most historians and locals treat it as a suburb—as do Robert B. Fairbanks and Philip G. Pope in this volume. Determining the exact suburban population of Texas is thus complicated.

As of 2000 the census bureau informed us that over 50 percent of the population of the United States resided in suburbs. Most Americans have also long shopped and worked in suburbs and spend an ever-increasing amount of their leisure and travel time moving from one suburb to the next. The bureau in 2010 listed 366 MSAs, up from 284 in 1990, 152 in 1940, ninety in 1910, twenty-nine in 1880, and nine in 1850. Again, an MSA encompasses suburbs outside a central city or twin cities of more than 50,000 people.[37]As of 2010, Texas had twenty-five MSAs. The largest in Texas, the Dallas–Fort Worth–Arlington area, was the fourth-largest in the nation with just shy of 6.4 million people (table 1). The second largest MSA in Texas, the Houston–Sugar Land–Baytown area, ranked fifth nationally (about six million). Texas also boasted three MSAs in the top twenty-five nationally (second to California's five), six in the top hundred, and five with over one million. As for individual cities, as of 2013, Texas had six in the top twenty nationally (the most of any state): Houston (fourth with about 2.2 million people); San Antonio (seventh with about 1.4 million); Dallas (ninth with about 1.25 million); Austin (eleventh with about 885,000); Fort Worth (seventeenth with about 800,000); and El Paso (nineteenth with about 675,000). As of 2010, Texas had the second-largest total urban population in the country (21,298,039). Another way to put it is that about 85 percent of Texas is so-called urban and almost 90 percent

TABLE 1. METROPOLITAN TEXAS, RANKINGS

TEXAS RANK	U.S. RANK	METROPOLITAN AREA	POPULATION
1	4	Dallas–Fort Worth–Arlington	6,371,773
2	5	Houston–Sugar Land–Baytown	6,086,538
3	25	San Antonio–New Braunfels	2,142,508
4	35	Austin–Round Rock–San Marcos	1,716,289
5	58	El Paso–Las Cruces	1,045,180
6	68	McAllen–Edinburg–Mission	774,769
7	114	Corpus Christi	442,600
8	126	Brownsville–Harlingen	406,220
9	127	Killeen–Temple–Fort Hood	405,300
10	132	Beaumont–Port Arthur	388,745
11	162	Lubbock	284,890
12	184	Laredo	250,304
13	185	Amarillo	249,881
14	188	Waco	234,906
15	192	College Station–Bryan	228,660
16	202	Tyler	209,714
17	198	Longview	206,874
18	240	Abilene	165,252
19	263	Wichita Falls	151,306
20	288	Texarkana	143,027
21	283	Odessa	137,130
22	284	Midland	136,872
23	313	Sherman–Denison	120,877
24	322	Victoria	115,384
25	326	San Angelo	111,823

Source: United States Census Bureau, American FactFinder, Population and Housing Occupancy Status: 2010—United States–Metropolitan Statistical Area; and for Puerto Rico 2010, Census National Summary File of Redistricting Data, available online at http://factfinder.census.gov.

TABLE 2. POPULATION CHANGE FOR METROPOLITAN STATISTICAL AREAS IN TEXAS, 1990–2014

METROPOLITAN STATISTICAL AREA	1990	2000	GROWTH RATE (%): 1990–2000	2010	GROWTH RATE (%): 2000–2010	2014	GROWTH RATE (%): 2010–2014	GROWTH RATE (%): 1990–2014
Abilene	148,004	160,245	8.3	165,252	3.1	168,592	2.0	13.9
Amarillo	196,144	226,522	15.5	249,881	10.3	259,885	4.0	32.5
Austin-Round Rock-San Marcos	846,227	1,249,763	47.7	1,716,289	37.3	1,943,299	13.2[a]	129.6
Beaumont-Port Arthur	361,226	385,090	6.6	388,745	0.9	405,427	1.0	12.2
Brownsville-Harlingen	260,120	335,227	28.9	406,220	21.2	420,392	3.5	61.6
College Station-Bryan	150,998	184,885	22.4	228,660	23.7	242,905	6.2	60.9
Corpus Christi	367,786	403,280	9.7	428,185	6.2	448,108	4.7	21.8
Dallas-Fort Worth-Arlington	3,989,294	5,161,544	29.4	6,371,773	23.4	6,954,330	9.1	74.3
El Paso	591,610	679,622	14.9	800,647	17.8	836,698	4.5	41.4
Houston-Sugar Land-Baytown	3,767,335	4,715,407	25.2	5,946,800	26.1	6,490,180	9.1[b]	72.3
Killeen-Temple-Fort Hood	268,822	330,714	23.0	405,300	22.6	424,858	4.8	58.0
Laredo	133,239	193,117	44.9	250,304	29.6	266,673	6.5	100.1
Longview	180,053	194,042	7.8	214,369	10.5	217,481	1.5	20.8
Lubbock	229,940	249,700	8.6	284,890	14.1	305,644	7.3	32.9
McAllen-Edinburg-Mission	383,545	569,463	48.5	774,769	36.1	831,073	7.3	116.7
Midland	106,611	116,009	8.8	136,872	18.0	161,290	17.8[c]	51.3
Odessa	118,934	121,123	1.8	137,130	13.2	153,904	12.2[d]	29.4
San Angelo	100,087	105,781	5.7	111,823	5.7	118,182	5.7	18.1
San Antonio-New Braunfels	1,407,745	1,711,703	21.6	2,142,508	25.2	2,328,652	8.7	65.4
Sherman-Denison	95,021	110,595	16.4	120,877	9.3	123,534	2.2	30.0
Texarkana, TX-Texarkana, AR	120,132	129,749	8.0	136,027	4.8	149,235	9.7	24.2
Tyler	151,309	174,706	15.5	209,714	20.0	218,842	4.4	44.6
Victoria	99,394	111,663	12.3	115,384	3.3	98,630	-14.5	-0.8
Waco	189,123	213,517	12.9	234,906	10.0	260,430	10.9	37.7
Wichita Falls	140,375	151,524	7.9	151,306	-0.1	151,536	0.2	8.0
Total	14,403,074	17,984,991	25	22,128,631	23.00	23,979,780	8.4	66.5
Texas	16,986,510	20,851,820	23	25,145,561	20.6	27,161,942	8	59.9

Source: Texas Department of State Health Services, Population Data for Texas, 1990, 200, 2010, and 2014.
[a] Fifth-highest growth rate among all national MSAs during 2010–2014 (Austin–Round Rock–San Marcos).
[b] Sixteenth-highest growth rate among all national MSAs during 2010–2014 Houston–Sugar Land–Baytown).
[c] Fourth-highest growth rate among all national MSAs during 2010–2014 (Midland).
[d] Sixth-highest growth rate among all national MSAs during 2010–2014 (Odessa).

is metropolitan. To put it bluntly, and in Texas slang, if you think Texas is rural, that dog won't hunt.

The census still allows us to get a snapshot of what is going on in Texas and to sound the alarm for scholars to come running. The Texas population data reveal that the MSA growth trend has been enormous for quite some time, dating back to World War II but markedly robust since 1990. The MSAs of Texas, collectively, have grown at a faster rate (67 percent) than the state overall (60 percent) (table 2). Texas boasted the fourth-, fifth-, sixth-, and sixteenth-fastest-growing metro areas in the United States from 2010 to 2014 (McAllen-Edinburg-Mission, Austin–Round Rock–San Marcos, Odessa, and Houston–Sugar Land–Baytown, respectively). Dallas–Fort Worth–Arlington and Houston–Sugar Land–Baytown added over 2 million inhabitants each since 1990, Austin–Round Rock–San Marcos added over a million, and San Antonio barely missed a million. In total, the MSAs in Texas added more than 9.5 million residents since 1990 while the entire state added just more than 10 million. Translation: almost all growth in Texas over the past thirty years has occurred in Texas's metropolitan areas.

Though metropolitan areas are unquestionably growing and have—for a long time—dominated Texas, specifically suburban data are harder to determine precisely because the census does not officially account for suburbs (and favors headcounts to, say, cultural or other markers of suburban distinction—discussed below). Moreover, as with Irving, census labels can mask previous "suburban" status in favor of current "urban" status. In fact, the story of many cities in Texas, from Irving to Plano to San Marcos and more, is correctly the story of urbanization by process of suburbanization. To put it differently, the urban history of Texas cannot be divorced from its suburban history.

Table 3 shows population growth data for the county seats of the so-called Triangle of Texas, a region that has also been called the Texaplex, as well as a "megapolitan" area and a "megaregion."[38] The three vertices of the so-called Texas Triangle—connected by Interstate Highways 45, 10, and 35—are the cities of Austin and San Antonio, Dallas and Fort Worth, and Houston (along with each of their surrounding counties). The table reveals the collective growth of these areas. Removing the population of the central cities from the total population, however, allows for comparison between the growth rate of urban and suburban areas.

Obviously some caveats are needed. The adjacent counties are not, in their entirety, suburban in population or character. Besides the "central

TABLE 3. POPULATION GROWTH IN TRIANGLE COUNTIES AND CITIES, 1940–2014

	1940	1990	GROWTH RATE (%): 1940–1990	2000	GROWTH RATE (%): 1990–2000	2010	GROWTH RATE (%): 2000–2010	2014	GROWTH RATE (%): 2010–2014	GROWTH RATE (%): 1940–2014	GROWTH RATE (%): 1990–2014
San Antonio–Austin counties	671,529	2,270,324	238	2,978,300	31	3,874,761	30	4,278,334	10	537	88
San Antonio city	253,854	935,933	269	1,144,646	22	1,327,407	16	1,416,291	7	458	51
Austin city	87,930	465,622	430	656,562	41	790,390	20	865,504	10	884	86
Cities combined	341,784	1,401,555	310	1,801,208	29	2,117,797	18	2,281,795	7	568	63
Counties without cities	329,745	868,769	164	1,177,092	36	1,756,964	49	1,996,539	14	506	130
Dallas–Fort Worth counties	916,381	3,987,814	335	5,154,237	29	6,339,387	23	6,870,232	8	650	72
Dallas city	294,734	1,006,977	242	1,188,580	18	1,197,816	1	1,241,162	4	321	23
Fort Worth city	177,662	447,619	152	534,697	19	741,206	39	781,000	5	340	74
Cities combined	472,396	1,454,596	208	1,723,277	18	1,939,022	13	2,022,162	4	328	39
Counties without cities	443,985	2,533,218	471	3,430,960	35	4,400,365	28	4,848,070	10	992	91
Houston counties	799,709	3,753,129	369	4,671,464	24	5,852,963	25	6,391,306	9	699	70
Houston city	384,514	1,630,553	324	1,953,631	20	2,100,263	8	2,233,310	6	481	37
Counties without city	415,195	2,122,576	411	2,717,833	28	3,752,700	38	4,157,996	11	901	96
Total county Growth for all areas	2,387,619	10,011,267	319	12,804,001	28	16,067,111	25	17,539,872	9	635	75
Total principal city growth combined	1,198,694	4,486,704	274	5,478,116	22	6,157,082	12	6,537,267	6	445	46
Total county growth without principal cities	1,200,964	5,524,563	360	7,325,885	33	9,910,029	35	11,002,605	11	816	99

Source: For county population numbers, see United States, Bureau of the Census, Sixteenth Census of the United States: 1940, Population, Vol. 2: Characteristics of the Population, Part 6: Pennsylvania-Texas, 792–806; Texas Department of State Health Services, Population Data for Texas, Texas Population by Area and County, 1990, 2000, 2010, and 2014. For city population numbers of Austin, Dallas, Fort Worth, Houston, and San Antonio, see United States, Bureau of the Census, Sixteenth Census of the United States: 1940, Population, Vol. 2: Characteristics of the Population, Part 6: Pennsylvania-Texas, 1010–53; United States, Bureau of the Census, 1990 Census of Population General Population Characteristics, Texas, Section 1 of 2; United States, Bureau of the Census, Population, Housing Units, Area, and Density: 2000 and 2010, State, Place, and (in selected states) County Subdivision, Census 2000 and 2010 Summary File 1 (SF 1) 100-Percent Data; and United States, Bureau of the Census, American Community Survey, Demographic and Housing Estimates, 2010–2014 American Community Survey 5-year Estimates (search "2014" for "Austin city, Texas" "Dallas city, Texas," "Fort Worth city, Texas," "Houston city, Texas," and "San Antonio city, Texas").

San Antonio and Austin counties: Atascosa, Bastrop, Bexar, Blanco, Burnet, Caldwell, Comal, Guadalupe, Hays, Lee, Medina, Travis, Williamson, and Wilson.

Dallas–Fort Worth counties: Collin, Dallas, Denton, Ellis, Hood, Hunt, Johnson, Kaufman, Parker, Tarrant, and Wise.

Houston counties: Brazoria, Chambers, Fort Bend, Harris, Jefferson, Liberty, Montgomery, and Waller.

cities," Triangle counties host one "city" with over 375,000 people (Arlington), three with over 200,000 (Plano, Garland, and Irving), and thirteen
with over 100,000 people. In addition, there are numerous small cities over
50,000 in size. Of course, rural areas certainly exist in these counties even
if their populations are small. In addition, within cities there exist posh,
suburb-like gated neighborhoods such as Greenway Parks in Dallas, Lan-
tana in Austin, and Carlton Woods in Houston. In fact, it is likely that the
overall data downplay actual suburban totals because of the high number of
so-called suburban urbanites residing within the borders of central cities,
as both Herb Ruffin II and Andrew Busch make clear about San Antonio
and Austin in this volume.

Since 1940 the county seats and surrounding counties of the Triangle
have collectively grown faster and larger than the central cities themselves.
The Triangle has collectively grown 635 percent. The central cities grew
445 percent (from 1,196,694 people in 1940 to 6,537,267 in 2014) while the
surrounding counties grew by 816 percent (from 1,200,964 to 11,002,605).
Only Austin and San Antonio grew more since 1940 than their host (Travis and Bexar) and surrounding counties, 568 percent to 506 percent. This
pattern of growth is consistent across the state. In the Dallas–Fort Worth
area the counties grew 992 percent since 1940 while Dallas and Fort Worth
cities, combined, 328 percent. In Houston the counties grew 901 percent
since 1940 and Houston city only 481 percent.

From 1990 to 2014 the counties of each area (again without the central
cities) grew more than the cities themselves (99 percent, collectively, for
counties to 46 percent, collectively, for the cities). If you consider the suburban areas of the Triangle alone, accepting all the caveats already mentioned, then the 11,002,605 residents of the suburban areas of the Triangle
constitute 41 percent of the state's population. In other words, about four
out of every ten Texans live in the suburbs of just the Triangle. Adding in
other metropolitan areas in Texas with suburbs, from El Paso to Corpus
Christi, it is clear that Texas is a majority suburban state.

When in the summer of 2014 it became known that cities throughout
the United States grew at a faster rate than suburbs between 2010 and 2013,
some people wrote of "the end of suburbia." So-called primary or central
cities—those over a million in this case—collectively grew 1.13 percent
compared to 0.95 percent growth in suburbs.[39] When considering a longer
time span, however, suburbia is instead thriving, especially in Texas. In fact,
between 2010 and the summer of 2012, eight of the fifteen fastest-growing

areas in the country were noncentral cities in Texas, such as Cedar Park, San Marcos, Georgetown, Frisco, McKinney, and Conroe. In addition, when considering both cities and suburbs, Texas had four of the top ten "Healthiest Housing Markets" in Austin, San Antonio, Fort Worth, and Houston. Said differently, both cities and suburbs in Texas grew, with suburbs outpacing cities, leaving rural areas to decline. For instance, between 2010 and 2012 Williamson County grew the fastest in the nation (7.94 percent), with Hays third (7.56 percent) and Fort Bend fifth (7.16 percent). Much of this came about because of booming tech industries like Dell in Round Rock, with a 73 percent increase in employment since 2000. Not to be outdone, Hays County experienced a 78 percent increase in employment since 2000 thanks to the expansion of energy companies outside Houston.[40] Ultimately, suburbs throughout the nation have rebounded, and the temporary phenomena of cities outpacing the growth of suburbs (never the case in Texas) likely happened because of the recession of 2008.[41] Surely, for cities to outpace suburbs in any statistically meaningful way, it would require decades or centuries for population numbers to catch up. For better or worse, then, long live suburbia, especially in Texas.

Indulge a deeper probe into city growth compared to suburban— so-called gentrification, which refers to "urban renewal" planning whereby wealthier (usually white) residents and (usually middle- to upper-class) businesses move in. This often results in driving out existing residents (often marked by differences in wealth and race) and causing property values to rise. This is the new white flight, they say: young, middle-class whites moving into "gentrified" city areas. This is all a part of the discussion of the growth of cities vis-à-vis suburbs and serves, for many, as a supposed explanation of why suburbia is losing dominance. Two things need consideration, however.

First, gentrification often follows a suburban aesthetic. This includes, according to scholar Greg Dickinson, a nostalgic longing for some mythic small-town past and lifestyle, emblematized by the architecture and design of so-called lifestyle centers and malls—"invented Main Streets."[42] Emphasizing navigable streets, historic building forms, multiuse neighborhoods, and pedestrian-friendly residential and retail landscapes, gentrification, it can be argued, is a thoroughly suburban form. Put differently, though urban because of city borders, gentrified areas are also aesthetically and culturally suburban, marking the phenomena as far more nuanced and complex than simple city versus suburb. Second, we need to stop favoring white people all the time, privileging where they are going and then naming the blessed phenomenon after them. Rather than just seeing white flight to cities, we are now seeing racially diverse flight (especially African American and Hispanic) to suburbs, so suburban flight does continue.

Appendix tables A through F chart growth and population numbers for the Triangle counties but with a focus on, first, the overall growth of each county since 1940 and, second, racial change and growth for each county since 1990. The growth in each county is indeed astounding. Since 1940 three Triangle counties grew by over 2,000 percent (Denton, Fort Bend, and Montgomery) and four grew over 1,000 percent (Brazoria, Collin, Hays, and Williamson). None of these counties hosted the larger cities in Texas. Overall, the Triangle counties have grown 635 percent since 1940, with Austin–San Antonio counties at 537 percent, Dallas–Fort Worth at 650 percent, and Houston at 699 percent. Considered altogether, 65 percent of all Texans lived in the Triangle in 2014, whereas only 37 percent did in 1940 (and, remember, most of that growth is in the suburban areas).

Some new caveats are needed as we turn to a discussion of race. Data concerning race are tricky. The census bureau has never been a pillar of consistency in establishing racial classifications and then tracking such data uniformly. This is due in part at least to past and present racism. Tracking African Americans has been consistent for over a century and a half, whereas white and Hispanic population numbers are blurred because Hispanics were, at times, listed as white or split between white head counts and "foreign born" from countries deemed Hispanic, most obviously Mexico. To confuse matters is the distinction between white Hispanics, stressing European heritage, and nonwhite Hispanics, leaving many Hispanics to identify themselves as "Other." In fact, after the Mexican-American War in 1848 and a desire to avoid extending arguments for citizenship and the rights thereof among nonwhites, between 1850 and 1920 the census bureau counted most Mexicans as racially "white." If this does not convince you of the social, cultural, and political construction of race, then nothing likely ever will. Given all this, to track racial demographics and change meaningfully at the state and county levels means looking over the past several decades only, most easily from 1990 forward.

Texas is a majority-minority state, with 2014 data telling us that the white population accounts for 43 percent of the state's total, African Americans 11 percent, Hispanics 40 percent, and "Other" 1 percent. Minorities constitute the majority of the population in the Triangle as well. The total population of the Triangle county areas, lumped together, including cities, was 17,539,872 residents by 2014. This included 7,346,619 identified as white (42 percent), 2,424,888 as African American (18 percent), 6,368,371 as Hispanic (36 percent), and 1,399,994 as the unfortunately titled "Other" (8 percent). Said otherwise, the Triangle grew in size but also grew more racially and culturally diverse as the nonwhite population accounts for about 58 percent of the region. Since 1990 the white population did grow, but at 18 percent compared to Hispanic growth of 201 percent, African American 72 percent, and "Other" 394 percent. Too, of the thirty-one Triangle counties in the appendix tables, whites were the absolute majority (over 50 percent) in twenty-nine of them in 1990. By 2014 whites were still dominant in most counties but lost absolute majority status in eight of them while Hispanics, who had held an absolute majority in only one county in 1990, enjoyed absolute majorities in three counties in 2014 (Atascosa, Bexar, and Medina—all in the Austin–San Antonio region). Also worth noting is that in 1990 whites held an absolute majority in each region (Austin–San Antonio, Dallas–Fort Worth, and Houston). By 2014 they no longer did and, in fact, fell behind Hispanics in both the Austin– San Antonio and the Houston regions. African Americans, while growing overall, decreased as a percentage of the population in many counties. For example, in 1990 African Americans were at least 10 percent of the population in eleven counties and two regions (Dallas–Fort Worth and Houston) and over 20 percent in Fort Bend and 30 percent in Jefferson and Waller. By 2014 they were at least 10 percent of the population in only six counties (but still both regions in Dallas–Fort Worth and Houston), 20 percent in three (Collin, Chambers, and Waller), and still 30 percent in Jefferson. Nevertheless, by 2014 the percentage of African Americans in a county's population increased in eighteen of the thirty-one counties listed. In other words, African American populations are growing throughout Texas and the Triangle areas, particularly in suburban counties, but not as much as Hispanics.

Ultimately, the population data in the appendix make clear that Texas suburbs are growing in both size and diversity, including immigrants. Indeed, though most students of history are familiar with the narrative of mass immigration to the nation's cities, the new narrative is that today's suburbia is the destination for most immigrants. The Brookings Institution, a nonprofit public policy organization out of Washington, D.C., reported that in 1990 the ten largest metropolitan areas with the largest amount of immigrant groups (including both the Dallas and Houston areas) accounted for 61 percent of all immigrants in the nation. By 2000 the number had dropped to 56 percent and by 2013 down to 51 percent. Still, in almost all metropolitan areas immigration totals increased between 2000 and 2013. Growth is still occurring, but in the suburbs rather than the central cities. By 2013, 61 percent of all immigrants in the United States lived in suburbs, up from 56 percent in 2000. Overall, 76 percent of the growth in immigrant population took place in the suburbs.[43]

Among African Americans a reverse Great Migration has taken place nationally such that 55 percent of all African Americans now live in the South. Notably, African Americans are migrating mostly to suburbs and not cities. In Texas, African American growth rates from 1990 to 2014 are higher in the surrounding counties near Bexar (57 percent), Travis (42 percent), Dallas (51 percent), Tarrant (110 percent), and Harris (50 percent). Around Bexar and Travis (homes to San Antonio and Austin cities) eight of twelve other counties had their African American population grow at an even faster rate than Bexar and Travis. In Austin city the African American population declined 5 percent, but in the Austin metropolitan area African Americans increased (i.e., they moved to suburbs). In Dallas–Fort Worth territory four of the other nine counties not named Dallas and Tarrant (homes of Dallas and Fort Worth cities) grew at a faster rate for African Americans than Dallas and Tarrant. In Houston three of the other seven counties not including Harris (home to Houston city) grew at higher rates for African Americans than Harris.[44]

A final word about minorities and population data generally: while minority populations are growing in suburbs nationally and in Texas, so is suburban poverty, with 16.5 million suburbanites reportedly living below the poverty line. The suburban poor are the largest poor population in the country.[45] The growth rate of the poor has risen higher in suburbs (139 percent) than cities (50 percent) since 2000. From 2000 to 2011 the growth of the suburban poor in the Austin metropolitan area (143 percent) made it second nationally to Atlanta (159 percent), with Dallas–Fort Worth–Arlington ranking twelfth (111 percent), Houston fifteenth (103 percent), and San Antonio fifty-fifth (53 percent).[46] The culprit seems to be gentrification. Urban planning has dislocated poor and minority populations thanks to new construction and subsequent rising property values. The result is that many of the poor and minority populations can no longer afford to live in the city.[47] Increasingly, they now call the suburbs home.

Unfortunately, most charitable and welfare organizations are still located in cities, and thus the suburban poor tend to lack the safety net and social services they had in the city.

SUBURBAN HISTORIOGRAPHY AND LONE STAR SUBURBS

Texas suburbs are booming, and it is time for historians to catch up with this trend. Suburban history, in fact, is an exciting and growing field of study. This is not surprising, since most Americans have lived there since the late twentieth century. As said, any understanding of the United States since World War II must account for suburbs. Moreover, suburban history can offer points of departure for Texas scholars to take new directions. This volume is intended to be such a departure point, to help suggest potential endeavors for others to pursue (or dispute/complicate). Ultimately, through suburban history, Texas can link to a national narrative rather than stay stuck on narrow parochial battles over Texas as southern or western. As historians Walter Buenger and Arnoldo De León explain, "Historians should not have to wrestle with the exasperating question of whether the state is Southern or Western, but should consider the state as a component of a larger enterprise, a component that fits into the nation and into the world in different ways at different times."48 Given that, we can contextualize and distinguish a Texan suburban experience that can very well be linked to multivariate currents, be they international, national, western, southern, European, Mexican, African American, and more.

In the United States suburbs have been studied and scrutinized since the colonial era, growing more robust by the time of the nineteenth-century romantic agrarianism of Catherine E. Beecher and Andrew Jackson Downing, then carrying on to late nineteenth-century criticism and condemnation of modernizing industrial society. Likewise, local, county, state, and regional histories in vogue throughout the nineteenth century, especially the work of Hubert Howe Bancroft and that of the Lewis Historical Publishing Company, explored suburban areas surrounding large cities. Although such publications usually amounted to local boosterism, meant to stimulate migration and investment by touting a celebratory past, they provided political, infrastructural, and biographical histories of some of suburbia's earliest days, particularly out west.49

In the first half of the twentieth century, especially in the wake of Frederick Jackson Turner, regional and sectional historians (such as Walter

Prescott Webb in Texas and the West) worked to document the mostly material and political forces shaping the growth, form, and functions of cities, suburbs, and outlying areas.[50] By the 1950s and 1960s sociologists and other professionals such as Herbert J. Gans and William H. Whyte had turned their attention to the shape of suburban life in a booming postwar America. Critics also continued to flourish, both before and after World War II, lamenting the by now usual suspects: alienation, uniformity, and loss of individuality in suburbia—that is, Little Boxes.[51]

Led by the rise of the "new social history" of the 1960s and 1970s that sought to look at society "from the bottom up," urban historians such as Stephan Thernstrom focused on processes that affected the lives of the urban masses and not just the elites. The resulting "new urban history," which sought to see the city as more than just a container for social action but also as a force affecting such activity, proved conducive to a more intense examination of suburbia.[52] Historian Sam Bass Warner Jr. had already begun looking at suburban development, form, and processes with his seminal 1962 publication of *Streetcar Suburbs: The Process of Growth in Boston, 1870–1900*, and Jon C. Teaford dissected fragmented metropolitan governments in his 1979 *City and Suburb: The Political Fragmentation of Metropolitan America, 1850–1970*. Warner and Teaford thus laid foundations for not only the suburban history that coalesced in the mid-1980s but also a more robust focus that sought to make the whole metropolitan area a unit of analysis.[53]

By the 1980s, works by John Archer, Henry C. Binford, Michael H. Ebner, Robert Fishman, Kenneth T. Jackson, Ann Durkin Keating, and Margaret Marsh, among others, ushered in fresh perspectives to the discussion of suburbia with analyses of demographics, technology, politics, social relationships, culture, gender, and the built environment. By the mid-1990s, a revisionist suburban history had arisen, particularly in the work of Margaret Crawford, Richard Harris, Becky M. Nicolaides, and Andrew Wiese. Often characterized as proponents of the "new suburban history," they argued against what they saw as a dominant representation of suburbia as demographically too simplistic (too white), geographically too narrow (too tied to the urban core), and functionally too particular (too residential). They lamented that the earlier focus on the white middle class glossed over such key phenomena as cultural, racial, and economic diversity (e.g., minority or blue-collar suburbs), varied housing stock (not just single-family homes), and the role of local communities and personalities.

The new suburban historians thus helped expand our understanding of suburbs by including examinations of more diverse peoples and places and focusing on their active production of the local.[54]

Since at least 2010 we are perhaps amid yet another "wave" or "turn" in suburban scholarship. In these works suburbia has once again been cast as highly varied and, as such, not a single space or type of space.[55] Taken together these works continue to undermine stereotypes about suburbia, particularly those concerning the built environment, form, and the cultural landscape, especially racial dynamics and minority lives. Indeed, these works powerfully shine a light on the continued reassembly and meaning of race in America, marking it not only as sociocultural but political, economic, and sadly human as well.

Another exciting dimension to this more recent wave of suburban studies is the intense focus on suburban agency, as well as the multilayered construction of meaning for and by suburbanites. This can be top-down, bottom-up, or, more often, an interactive and co-constructivist dynamic of top-down *and* bottom-up. Be it the creativity of suburbanites themselves; the structured and structuring mold and effects of broader federal, state, and local laws, narratives, and beliefs; or some dynamic blending of it all, suburbanites and their activity—their making, their living, their performing, and their dreaming—are a major thrust of analysis and consideration in recent suburban history.

As John Archer has argued, the focus here is on the continual reproduction and reassembling of suburbia as an artifact and aspect of people's lives, emphasizing that suburbs need suburbanites, with their own agendas, to exist.[56] Certainly, attention goes to the conditions and circumstances informing the production process, but as theorist Michel de Certeau argued, people can "poach" and engage their environment in their everyday lives as much as they engage and use material objects.[57] People work within their own environments, using what is available to them to make their own lives and fashion their own understandings.[58] In other words, society, including places such as suburbs, is a context—a material, social, and discursive apparatus that is malleable. Inherited or lived in, conditions and circumstances are a menu of both opportunities and limitations. Still, as Archer again argues, people are not prisoners of the menu, since human creativity and an endless variety of menu items generate diversity. Suburbia, then, is indeed both an artifact and an aspect of people's lives, allowing scholars to approach suburban lives as multifaceted and complex, returning a role of agency and activity to suburbanites in the fashioning of their own lives—"to follow," as theorist Bruno Latour once pleaded, "the actors themselves, that is try to catch up with their often wild innovations in order to learn from them what the collective existence has become in their hands."[59]

This volume looks to both past and recent trends in suburban history. Just as country singer Kitty Wells reminded us that "it wasn't God who made honky tonk angels," cities have not always made their suburbs in Texas. But neither do Texas suburbs just spring forth. They are achieved. Although of course cities have been a part of the equation, it has been suburbanites themselves who have worked to make their suburbs. Suburbs, even in Texas, resulted from the decisions of particular people in particular contexts and in particular places. Whether it was transportation, growth desires, and the hopes of making a buck, or the result of legislation, migration trends, capital flow, or environmental concerns, Lone Star suburbs were indeed made—they did not just "appear." There was a specific process and one that intentionally produced the suburban culture that constitutes the majority of Texas in the twenty-first century.

Authors in this volume thus look at the various factors that shaped Texas suburbs while anchoring them to particular contexts and naming names. Case studies range from explorations of how suburban leaders took the reins in planning and the use of government policies, such as in Irving (Philip G. Pope), North Dallas (Robert B. Fairbanks), and outside Houston (Andrew C. Baker), to how suburbanites of various ethno-racial groups, such as African Americans (Herb Ruffin II on San Antonio and Ted and Gail Lawe about Dallas) and Asian Americas (Son Mai), looked to make suburbia their own. Tom McKinney reviews the construction of highways that made suburbia possible in Texas while also highlighting the role of place-based leaders making use of federal and state laws. One effect of highways and reliance on automobiles has been a concern for the environment. Andrew Busch thus tackles the rise of environmentalism in Austin's suburbs, taking care to note the role of suburbanites in fashioning Austin and its suburbs in their image. Jake McAdams concerns himself with image and representation, exploring the rise of the "cowboy church" in Texas as a suburban phenomenon.

From all these contributions flows the realization that suburbs, especially Lone Star suburbs, are not exclusively white, nor are they exclusively residential. Increasingly, Texas suburbia means diversity of race and diversity of form and function. Additionally, the authors highlight the role of creative individuals within local contexts, motivated at times by profit but also by the impact of local, state, and federal apparatuses representative of (neo)liberal pro-growth agendas. The contributors examine elements ranging from the built environment, such as freeways and architecture, to government policies, to the role of site-centered elites and entrepreneurs actively looking to grow their suburbs. The collective result gives us a look under the hood to see how suburbs in Texas take shape, how they grow, and why. Often deplored for its ill effects, suburbia is a cultural landscape in which individual Texans and groups make and live their lives. The dynamics of race, class, gender, perception, and sociocultural pressures play out in real time and in real places. Suburbia, of course, is such a place. Suburban landscapes mix with other pressures and structures to affect the daily lives of Texas suburbanites who forge identity, meaning, and aspirations within the suburbs they live. The authors in this volume provide fresh perspectives to highlight how Texans have sought to make the Lone Star suburbs their own.

APPENDIX: TRIANGLE OF TEXAS POPULATION DATA
TABLE A. SELECTED COUNTY POPULATION OF TRIANGLE AREA, 1940

	TOTAL	TOTAL (%)
TEXAS	6,414,824	100.00
METROPOLITAN	2,911,389	45.39
AUSTIN AND SAN ANTONIO COUNTIES		
Atascosa	19,275	0.30
Bastrop	21,610	0.34
Bexar	338,176	5.27
Blanco	4,264	0.07
Burnet	10,771	0.17
Caldwell	24,893	0.39
Comal	12,321	0.19
Guadalupe	25,596	0.40
Hays	15,349	0.24
Lee	12,751	0.20
Medina	16,106	0.25
Travis	111,653	1.74
Williamson	41,698	0.65
Wilson	17,066	0.27
Total	**671,529**	**10.47**
DALLAS—FORT WORTH COUNTIES		
Collin	47,190	0.74
Dallas	398,564	6.21
Denton	33,658	0.52
Ellis	47,733	0.74
Hood	6,674	0.10
Hunt	48,793	0.76
Johnson	30,384	0.47
Kaufman	38,308	0.60
Parker	20,482	0.32
Tarrant	225,521	3.52
Wise	19,074	0.30
Total	**916,381**	**14.29**
HOUSTON COUNTIES		
Brazoria	27,069	0.42
Chambers	7,511	0.12
Fort Bend	32,963	0.51
Harris	528,961	8.25
Jefferson	145,329	2.27
Liberty	24,541	0.38
Montgomery	23,055	0.36
Waller	10,280	0.16
Total	**799,709**	**12.47**
Grand Total	**2,387,619**	**37.22**

Source: United States, Bureau of the Census, *Sixteenth Census of the United States: 1940, Population, Vol. 2: Characteristics of the Population, Part 6: Pennsylvania-Texas,* 792–806.

TABLE B. SELECTED COUNTY POPULATION OF TRIANGLE AREA, 1990

	TOTAL	% OF TOTAL	GROWTH RATE (%): 1940–1990	WHITE	WHITE (%)
TEXAS	16,986,510	100.00	165	10,308,444	60.69
METROPOLITAN	14,165,650	83.00	387	8,399,249	59.29
AUSTIN AND SAN ANTONIO COUNTIES					
Atascosa	30,533	0.18	58	14,213	46.55
Bastrop	38,263	0.23	77	26,686	69.74
Bexar	1,185,394	6.98	251	497,074	41.93
Blanco	5,972	0.04	40	5,041	84.41
Burnet	22,677	0.13	111	19,825	87.42
Caldwell	26,392	0.16	6	13,562	51.39
Comal	51,832	0.31	321	39,262	75.75
Guadalupe	64,873	0.38	153	41,503	63.98
Hays	65,614	0.39	327	44,708	68.14
Lee	12,854	0.08	1	9,654	75.11
Medina	27,312	0.16	70	14,946	54.72
Travis	576,407	3.39	416	376,111	65.25
Williamson	139,551	0.82	235	110,854	79.44
Wilson	22,650	0.13	33	14,286	63.07
Total	2,270,324	13.37	238	1,227,725	54.08
DALLAS–FORT WORTH COUNTIES					
Collin	264,036	1.55	460	227,100	86.01
Dallas	1,852,810	10.91	365	1,117,475	60.31
Denton	273,525	1.61	713	233,319	85.30
Ellis	85,167	0.50	78	64,934	76.24
Hood	28,981	0.17	334	27,272	94.10
Hunt	64,343	0.38	32	54,150	84.16
Johnson	97,165	0.57	220	86,487	89.01
Kaufman	52,220	0.31	36	41,246	78.99
Parker	64,785	0.38	216	60,986	94.14
Tarrant	1,170,103	6.89	419	858,901	73.40
Wise	34,679	0.20	82	31,360	90.43
Total	3,987,814	23.48	335	2,803,230	70.29
HOUSTON COUNTIES					
Brazoria	191,707	1.13	608	139,841	72.95
Chambers	20,088	0.12	167	16,181	80.55
Fort Bend	225,421	1.33	584	121,737	54.00
Harris	2,818,199	16.59	433	1,532,412	54.38
Jefferson	239,397	1.41	65	147,534	61.63
Liberty	52,726	0.31	115	42,717	81.02
Montgomery	182,201	1.07	690	159,549	87.57
Waller	23,390	0.14	128	11,961	51.14
Total	3,753,129	22.09	369	2,171,932	57.87
Grand Total	10,011,267	58.94	320	6,202,887	61.96
% of state's selected population	n/a	n/a	n/a	60	n/a

Source: Texas Department of State Health Services, Population Data for Texas, Texas Population by Area and County, 1990.

AFRICAN AMERICAN	AFRICAN AMERICAN (%)	HISPANIC	HISPANIC (%)	OTHER	OTHER (%)
1,980,693	11.66	4,339,900	25.55	357,473	2.10
1,744,834	12.32	3,684,315	26.01	337,260	2.38
108	0.35	16,054	52.58	158	0.52
4,356	11.38	6,925	18.10	296	0.77
81,533	6.88	589,123	49.70	17,664	1.49
56	0.94	839	14.05	36	0.60
255	1.12	2,438	10.75	159	0.70
2,678	10.15	9,980	37.81	172	0.65
421	0.81	11,852	22.87	297	0.57
3,456	‹5.33	19,233	29.65	681	1.05
2,094	3.19	18,234	27.79	578	0.88
1,753	13.64	1,408	10.95	39	0.30
71	0.26	12,126	44.40	169	0.62
61,152	10.61	121,731	21.12	17,413	3.02
6,548	4.69	19,992	14.33	2,157	1.55
235	1.04	8,046	35.52	83	0.37
164,716	7.26	837,981	36.91	39,902	1.76
10,752	4.07	18,161	6.88	8,023	3.04
363,015	19.59	315,710	17.04	56,610	3.06
13,344	4.88	19,014	6.95	7,848	2.87
8,416	9.88	11,230	13.19	587	0.69
52	0.18	1,351	4.66	306	1.06
6,743	10.48	2,872	4.46	578	0.90
2,452	2.52	7,448	7.67	778	0.80
7,240	13.86	3,336	6.39	398	0.76
570	0.88	2,693	4.16	536	0.83
138,608	11.85	139,886	11.96	32,708	2.80
387	1.12	2,660	7.67	272	0.78
551,579	13.83	524,361	13.15	108,644	2.72
15,449	8.06	33,774	17.62	2,643	1.38
2,542	12.65	1,193	5.94	172	0.86
45,879	20.35	43,987	19.51	13,818	6.13
529,572	18.79	645,504	22.90	110,711	3.93
73,903	30.87	12,623	5.27	5,337	2.23
6,826	12.95	2,876	5.45	307	0.58
7,668	4.21	13,222	7.26	1,762	0.97
8,743	37.38	2,588	11.06	98	0.42
690,582	18.40	755,767	20.14	134,848	3.59
1,406,877	14.05	2,118,109	21.16	283,394	2.83
71	n/a	49	n/a	79	n/a

TABLE C. SELECTED COUNTY POPULATION OF TRIANGLE AREA, 2000

	TOTAL	% OF TOTAL	GROWTH RATE (%): 1990–2000	WHITE	WHITE (%)
TEXAS	20,851,820	100.00	23	11,074,716	53.11
METROPOLITAN	17,691,880	84.85	25	9,066,062	51.24
AUSTIN AND SAN ANTONIO COUNTIES					
Atascosa	38,628	0.19	27	15,488	40.10
Bastrop	57,733	0.28	51	38,169	66.11
Bexar	1,392,931	6.68	18	506,093	36.33
Blanco	8,418	0.04	41	7,000	83.16
Burnet	34,147	0.16	51	28,258	82.75
Caldwell	32,194	0.15	22	16,149	50.16
Comal	78,021	0.37	51	58,871	75.46
Guadalupe	89,023	0.43	37	53,601	60.21
Hays	97,589	0.47	49	63,684	65.26
Lee	15,657	0.08	22	10,797	68.96
Medina	39,304	0.19	44	20,214	51.43
Travis	812,280	3.90	41	465,317	57.29
Williamson	249,967	1.20	79	185,788	74.33
Wilson	32,408	0.16	43	19,936	61.52
Total	2,978,300	14.28	31	1,489,365	50.01
DALLAS–FORT WORTH COUNTIES					
Collin	491,675	2.36	86	379,088	77.10
Dallas	2,218,899	10.64	20	998,543	45.00
Denton	432,976	2.08	58	333,058	76.92
Ellis	111,360	0.53	31	80,115	71.94
Hood	41,100	0.20	42	37,492	91.22
Hunt	76,596	0.37	19	61,738	80.60
Johnson	126,811	0.61	31	106,395	83.90
Kaufman	71,313	0.34	37	54,886	76.96
Parker	88,495	0.42	37	79,689	90.05
Tarrant	1,446,219	6.94	24	908,197	62.80
Wise	48,793	0.23	41	42,395	86.89
Total	5,154,237	24.72	29	3,081,596	59.79
HOUSTON COUNTIES					
Brazoria	241,767	1.16	26	159,797	66.10
Chambers	26,031	0.12	30	20,339	78.13
Fort Bend	354,452	1.70	57	166,758	47.05
Harris	3,400,578	16.31	21	1,456,811	42.84
Jefferson	252,051	1.21	5	131,753	52.27
Liberty	70,154	0.34	33	52,734	75.17
Montgomery	293,768	1.41	61	241,180	82.10
Waller	32,663	0.16	40	16,447	50.35
Total	4,671,464	22.40	24	2,245,819	48.08
Grand Total	12,804,001	61.40	28	6,816,780	53.24
% of state's selected population	n/a	n/a	n/a	62	n/a

Source: Texas Department of State Health Services, Population Data for Texas, Texas Population by Area and County, 2000.

AFRICAN AMERICAN	AFRICAN AMERICAN (%)	HISPANIC	HISPANIC (%)	OTHER	OTHER (%)
2,421,653	11.61	6,669,666	31.99	685,785	3.29
2,159,135	12.20	5,809,786	32.84	656,897	3.71
207	0.54	22,620	58.56	313	0.81
5,110	8.85	13,845	23.98	609	1.05
100,260	7.20	757,033	54.35	29,545	2.12
71	0.84	1,290	15.32	57	0.68
551	1.61	5,044	14.77	294	0.86
2,761	8.58	13,018	40.44	266	0.83
791	1.01	17,609	22.57	750	0.96
4,541	5.10	29,561	33.21	1,320	1.48
3,653	3.74	28,859	29.57	1,393	1.43
1,902	12.15	2,848	18.19	110	0.70
838	2.13	17,873	45.47	379	0.96
76,192	9.38	229,048	28.20	41,723	5.14
13,185	5.27	42,990	17.20	8,004	3.20
393	1.21	11,834	36.52	245	0.76
210,455	7.07	1,193,472	40.07	85,008	2.85
24,509	4.98	50,510	10.27	37,568	7.64
454,103	20.47	662,729	29.87	103,524	4.67
26,290	6.07	52,619	12.15	21,009	4.85
9,725	8.73	20,508	18.42	1,012	0.91
150	0.36	2,975	7.24	483	1.18
7,410	9.67	6,366	8.31	1,082	1.41
3,274	2.58	15,375	12.12	1,767	1.39
7,674	10.76	7,925	11.11	828	1.16
1,657	1.87	6,211	7.02	938	1.06
188,144	13.01	285,290	19.73	64,588	4.47
648	1.33	5,248	10.76	502	1.03
723,584	14.04	1,115,756	21.65	233,301	4.53
20,747	8.58	55,063	22.78	6,160	2.55
2,573	9.88	2,810	10.79	309	1.19
70,810	19.98	74,871	21.12	42,013	11.85
630,184	18.53	1,119,751	32.93	193,832	5.70
85,267	33.83	26,536	10.53	8,495	3.37
9,115	12.99	7,660	10.92	645	0.92
10,481	3.57	37,150	12.65	4,957	1.69
9,603	29.40	6,344	19.42	269	0.82
838,780	17.96	1,330,185	28.47	256,680	5.49
1,772,819	13.85	3,639,413	28.42	574,989	4.49
73	n/a	55	n/a	84	n/a

TABLE D. SELECTED COUNTY POPULATION OF TRIANGLE AREA, 2010

	TOTAL	% OF TOTAL	GROWTH RATE (%): 1990–2000	WHITE	WHITE (%)
TEXAS	25,145,561	100.00	21	11,397,345	45.33
METROPOLITAN	22,140,398	88.05	25	9,637,077	43.53
AUSTIN AND SAN ANTONIO COUNTIES					
Atascosa	44,911	0.18	16	16,295	36.28
Bastrop	74,171	0.29	28	42,446	57.23
Bexar	1,714,773	6.82	23	519,123	30.27
Blanco	10,497	0.04	25	8,336	79.41
Burnet	42,750	0.17	25	32,530	76.09
Caldwell	38,066	0.15	18	16,841	44.24
Comal	108,472	0.43	39	77,387	71.34
Guadalupe	131,533	0.52	48	72,086	54.80
Hays	157,107	0.62	61	92,062	58.60
Lee	16,612	0.07	6	10,798	65.00
Medina	46,006	0.18	17	21,408	46.53
Travis	1,024,266	4.07	26	517,644	50.54
Williamson	422,679	1.68	69	269,481	63.76
Wilson	42,918	0.17	32	25,186	58.68
Total	3,874,761	15.41	30	1,721,623	44.43
DALLAS–FORT WORTH COUNTIES					
Collin	782,341	3.11	59	493,492	63.08
Dallas	2,368,139	9.42	7	784,693	33.14
Denton	662,614	2.64	53	426,887	64.42
Ellis	149,610	0.59	34	97,987	65.49
Hood	51,182	0.20	25	44,588	87.12
Hunt	86,129	0.34	12	64,393	74.76
Johnson	150,934	0.60	19	115,545	76.55
Kaufman	103,350	0.41	45	72,328	69.98
Parker	116,927	0.47	32	99,698	85.27
Tarrant	1,809,034	7.19	25	937,135	51.80
Wise	59,127	0.24	21	47,122	79.70
Total	6,339,387	25.21	23	3,183,868	50.22
HOUSTON COUNTIES					
Brazoria	313,166	1.25	30	166,674	53.22
Chambers	35,096	0.14	35	24,767	70.57
Fort Bend	585,375	2.33	65	211,680	36.16
Harris	4,092,459	16.28	20	1,349,646	32.98
Jefferson	252,273	1.00	0	112,503	44.60
Liberty	75,643	0.30	8	52,321	69.17
Montgomery	455,746	1.81	55	324,611	71.23
Waller	43,205	0.17	32	19,260	44.58
Total	5,852,963	23.28	25	2,261,462	38.64
Grand Total	16,067,111	63.90	25	7,166,953	44.61
% of state's selected population	n/a	n/a	n/a	63	n/a

Source: Texas Department of State Health Services, Population Data for Texas, Texas Population by Area and County, 2010.

AFRICAN AMERICAN	AFRICAN AMERICAN (%)	HISPANIC	HISPANIC (%)	OTHER	OTHER (%)
2,886,825	11.48	9,460,921	37.62	1,400,470	5.57
2,651,414	11.98	8,511,794	38.44	1,340,113	6.05
256	0.57	27,785	61.87	575	1.28
5,536	7.46	24,190	32.61	2,000	2.70
118,460	6.91	1,006,958	58.72	70,232	4.10
62	0.59	1,909	18.19	190	1.81
700	1.64	8,652	20.24	868	2.03
2,456	6.45	17,922	47.08	847	2.23
1,606	1.48	26,989	24.88	2,490	2.30
7,963	6.05	46,889	35.65	4,595	3.49
4,970	3.16	55,401	35.26	4,674	2.98
1,772	10.67	3,724	22.42	318	1.91
913	1.98	22,871	49.71	814	1.77
82,805	8.08	342,766	33.46	81,051	7.91
24,744	5.85	98,034	23.19	30,420	7.20
644	1.50	16,412	38.24	676	1.58
252,886	6.53	1,700,502	43.89	199,750	5.16
64,715	8.27	115,354	14.74	108,780	13.90
518,732	21.90	905,940	38.26	158,774	6.70
54,034	8.15	120,836	18.24	60,857	9.18
13,161	8.80	35,161	23.50	3,301	2.21
225	0.44	5,234	10.23	1,135	2.22
6,976	8.10	11,751	13.64	3,009	3.49
3,797	2.52	27,319	18.10	4,273	2.83
10,571	10.23	17,548	16.98	2,903	2.81
1,842	1.58	12,410	10.61	2,977	2.55
262,522	14.51	482,977	26.70	126,400	6.99
573	0.97	10,112	17.10	1,320	2.23
937,148	14.78	1,744,642	27.52	473,729	7.47
36,880	11.78	86,643	27.67	22,969	7.33
2,817	8.03	6,635	18.91	877	2.50
123,267	21.06	138,967	23.74	111,461	19.04
754,258	18.43	1,671,540	40.84	317,015	7.75
84,500	33.50	42,889	17.00	12,371	4.90
8,074	10.67	13,602	17.98	1,646	2.18
18,537	4.07	94,698	20.78	17,900	3.93
10,537	24.39	12,536	29.02	872	2.02
1,038,870	17.75	2,067,520	35.32	485,111	8.29
2,228,904	13.87	5,512,664	34.31	1,158,590	7.21
77	n/a	58	n/a	83	n/a

TABLE E. SELECTED COUNTY POPULATION OF TRIANGLE AREA, 2014

	TOTAL	% OF TOTAL	GROWTH RATE (%): 2010–2014	WHITE	WHITE (%)
TEXAS	27,161,942	100.00	8	11,624,881	42.80
METROPOLITAN	24,027,285	88.46	9	9,841,287	40.96
AUSTIN AND SAN ANTONIO COUNTIES					
Atascosa	49,165	0.18	9	17,195	34.97
Bastrop	83,586	0.31	13	44,729	53.51
Bexar	1,847,931	6.80	8	521,326	28.21
Blanco	11,478	0.04	9	9,024	78.62
Burnet	46,398	0.17	9	34,395	74.13
Caldwell	42,215	0.16	11	17,739	42.02
Comal	124,179	0.46	14	87,447	70.42
Guadalupe	150,518	0.55	14	79,668	52.93
Hays	188,705	0.69	20	106,777	56.58
Lee	17,624	0.06	6	11,199	63.54
Medina	50,202	0.18	9	22,747	45.31
Travis	1,119,822	4.12	9	545,929	48.75
Williamson	498,102	1.83	18	302,647	60.76
Wilson	48,409	0.18	13	28,015	57.87
Total	4,278,334	15.75	10	1,828,837	42.75
DALLAS–FORT WORTH COUNTIES					
Collin	913,737	3.36	17	541,132	59.22
Dallas	2,469,911	9.09	4	721,025	29.19
Denton	762,161	2.81	15	460,118	60.37
Ellis	168,296	0.62	12	105,512	62.69
Hood	55,084	0.20	8	47,343	85.95
Hunt	91,849	0.34	7	66,038	71.90
Johnson	164,246	0.60	9	120,499	73.36
Kaufman	119,752	0.44	16	80,468	67.20
Parker	132,345	0.49	13	109,829	82.99
Tarrant	1,928,056	7.10	7	927,003	48.08
Wise	64,795	0.24	10	49,671	76.66
Total	6,870,232	25.29	8	3,228,638	46.99
HOUSTON COUNTIES					
Brazoria	349,214	1.29	12	172,811	49.49
Chambers	39,739	0.15	13	26,866	67.61
Fort Bend	694,429	2.56	19	230,669	33.22
Harris	4,391,445	16.17	7	1,318,226	30.02
Jefferson	257,872	0.95	2	107,286	41.60
Liberty	81,483	0.30	8	54,441	66.81
Montgomery	528,509	1.95	16	358,075	67.75
Waller	48,615	0.18	13	20,770	42.72
Total	6,391,306	23.53	9	2,289,144	35.82
Grand Total	17,539,872	64.58	9	7,346,619	41.89
% of state's selected population	n/a	n/a	n/a	63	n/a

Source: Texas Department of State Health Services, Population Data for Texas, Texas Population by Area and County, 2014.

AFRICAN AMERICAN	AFRICAN AMERICAN (%)	HISPANIC	HISPANIC (%)	OTHER	OTHER (%)
3,114,187	11.47	10,740,456	39.54	1,682,418	6.19
2,870,171	11.95	9,700,948	40.37	1,614,879	6.72
273	0.56	31,064	63.18	633	1.29
6,151	7.36	30,236	36.17	2,470	2.96
128,356	6.95	1,112,464	60.20	85,785	4.64
60	0.52	2,189	19.07	205	1.79
775	1.67	10,266	22.13	962	2.07
2,683	6.36	20,849	49.39	944	2.24
1,960	1.58	31,609	25.45	3,163	2.55
9,314	6.19	55,783	37.06	5,753	3.82
5,777	3.06	70,129	37.16	6,022	3.19
1,883	10.68	4,188	23.76	354	2.01
945	1.88	25,598	50.99	912	1.82
86,913	7.76	389,682	34.80	97,298	8.69
30,352	6.09	126,268	25.35	38,835	7.80
691	1.43	18,879	39.00	824	1.70
276,133	6.45	1,929,204	45.09	244,160	5.71
82,217	9.00	150,942	16.52	139,446	15.26
548,478	22.21	1,017,542	41.20	182,866	7.40
66,088	8.67	156,304	20.51	79,651	10.45
15,266	9.07	43,364	25.77	4,154	2.47
236	0.43	6,229	11.31	1,276	2.32
7,627	8.30	14,448	15.73	3,736	4.07
4,500	2.74	33,947	20.67	5,300	3.23
12,484	10.42	23,127	19.31	3,673	3.07
2,086	1.58	16,707	12.62	3,723	2.81
290,535	15.07	561,809	29.14	148,709	7.71
608	0.94	12,926	19.95	1,590	2.45
1,030,125	14.99	2,037,345	29.65	574,124	8.36
44,086	12.62	102,628	29.39	29,689	8.50
3,220	8.10	8,608	21.66	1,045	2.63
147,484	21.24	174,903	25.19	141,373	20.36
795,556	18.12	1,907,830	43.44	369,833	8.42
86,114	33.39	50,298	19.51	14,174	5.50
8,556	10.50	16,579	20.35	1,907	2.34
22,406	4.24	125,331	23.71	22,697	4.29
11,208	23.05	15,645	32.18	992	2.04
1,118,630	17.50	2,401,822	37.58	581,710	9.10
2,424,888	13.83	6,368,371	36.31	1,399,994	7.98
78	n/a	59	n/a	83	n/a

TABLE F. GROWTH RATE OF SELECTED POPULATIONS, 1990–2014

	GROWTH RATE (%): TOTAL POPULATION, 1940–2014	GROWTH RATE (%), 1990–2014				
		TOTAL	WHITE	AFRICAN AMERICAN	HISPANIC	OTHER
TEXAS	323	60	13	57	147	371
METROPOLITAN	725	70	17	64	163	379
AUSTIN AND SAN ANTONIO COUNTIES						
Atascosa	155	61	21	153	93	301
Bastrop	287	118	68	41	337	734
Bexar	446	56	5	57	89	386
Blanco	169	92	79	7	161	469
Burnet	331	105	73	204	321	505
Caldwell	70	60	31	0	109	449
Comal	908	140	123	366	167	965
Guadalupe	488	132	92	170	190	745
Hays	1129	188	139	176	285	942
Lee	38	37	16	7	197	808
Medina	212	84	52	1231	111	440
Travis	903	94	45	42	220	459
Williamson	1095	257	173	364	532	1700
Wilson	184	114	96	194	135	893
Total	537	88	49	68	130	512
DALLAS–FORT WORTH COUNTIES						
Collin	1836	246	138	665	131	1638
Dallas	520	33	-35	51	222	223
Denton	2164	179	97	395	722	915
Ellis	253	98	62	81	286	608
Hood	725	90	74	354	361	317
Hunt	88	43	22	13	403	546
Johnson	441	69	39	84	356	581
Kaufman	213	129	95	72	593	823
Parker	546	104	80	266	520	595
Tarrant	755	65	8	110	302	355
Wise	240	87	58	57	386	485
Total	650	72	15	87	289	428
HOUSTON COUNTIES						
Brazoria	1190	82	24	185	204	1023
Chambers	429	98	66	27	622	508
Fort Bend	2007	208	89	221	298	923
Harris	730	56	-14	50	196	234
Jefferson	77	8	-27	17	298	166
Liberty	232	55	27	25	476	521
Montgomery	2192	190	124	192	848	1186
Waller	373	108	74	28	505	912
Total	699	70	5	62	218	331
GrandTotal	635	75	18	72	201	394

Source: Created by author based on Tables A–E.

38

NOTES

1. For similar examples of historians lamenting this old debate, see contributions by Alwyn Barr, Robert A. Calvert, Fane Downs, Frank Vandiver, and Randolph B. Campbell in Walter L. Buenger and Robert A. Calvert, eds., *Texas through Time: Evolving Interpretations* (College Station: Texas A&M University Press, 1991).

2. Calvert, "Agrarian Texas," in Buenger and Calvert, *Texas through Time*, 198.

3. For examples, see Michael E. Young, "Census: Suburbs, Oil and Gas Fuel Texas Population Growth," *Dallas Morning News*, May 22, 2013; Quinton Renfro, "Texas Dominates List of Top High Growth Suburban Areas," January 12, 2011, https://activerain.com/blogsview/2072200/texas-dominates-list-of-top-10-high-growth-suburban-areas; Lomi Kriel and Dug Begley, "Census Data Show [Houston] City Outgained All but NYC, while Lagging County," *Houston Chronicle*, May 21, 2014; Joel Kotkin, "America's Fastest-Growing Counties: The 'Burbs Are Back," *Forbes*, September 26, 2013, www.forbes.com/sites/joelkotkin/2013/09/26/americas-fastest-growing-counties-the-burbs-are-back; Wendell Cox, "Texas Suburbs Lead Population Growth," *New Geography*, May 25, 2013; and Alexa Ura and Chris Essing, "Harris County Loses Residents to Other Areas; Texas Suburbs Growing," *Texas Tribune*, March 23, 2017.

4. See Light Townsend Cummins and Alvin R. Bailey, *A Guide to the History of Texas* (New York: Greenwood Press, 1988); Buenger and Calvert, *Texas through Time*; Sam W. Haynes, ed., *Major Problems in Texas History* (Boston: Mifflin, 2000); Laura Lyons McLemore, *Inventing Texas: Early Historians of the Lone Star State* (College Station: Texas A&M University Press, 2004); Richard B. McCaslin, *At the Heart of Texas: One Hundred Years of the Texas State Historical Association, 1897–1997* (Austin: Texas State Historical Association, 2007); Gregg Cantrell and Elizabeth Hayes Turner, eds., *Lone Star Pasts: Memory and History in Texas* (College Station: Texas A&M University Press, 2006); Walter L. Buenger and Arnoldo De León, eds., *Beyond Texas through Time: Breaking Away from Past Interpretations* (College Station: Texas A&M University Press, 2011); Bruce A. Glasrud, Light Townsend Cummins, and Cary D. Wintz, eds., *Discovering Texas History* (Norman: University of Oklahoma Press, 2014); and Light Townsend Cummins and Mary L. Scheer, *Texan Identities: Moving beyond Myth, Memory, and Fallacy in Texas History* (Denton: University of North Texas Press, 2016).

5. For just a few recent examples, see Linda Scarbrough, *Road, River, and Ol' Boy Politics: A Texas County's Path from Farm to Supersuburb* (Austin: Texas State Historical Association, 2005); Patrick G. Williams, *Beyond Redemption: Texas Democrats after Reconstruction* (College Station: Texas A&M Press, 2007); Kyle G. Wilkinson, *Yoeman, Sharecroppers, and Socialists: Plain Folk Protest in Texas, 1870–1914* (College Station: Texas A&M Press, 2008); David O'Donald Cullen and Kyle G. Wilkison, eds., *The Texas Left: The Radical Roots of Lone Star Liberalism* (College Station: Texas A&M University Press, 2010); Sean P. Cunningham, *Cowboy Conservatism: Texas and the Rise of the Modern Right* (Lexington: University Press of Kentucky, 2010); and Wayne Thorburn, *Red State: An Insider's Story of How the GOP Came to Dominate Texas Politics* (Austin: University of Texas Press, 2014).

6. Alun Munslow, *Narrative and History* (New York: Palgrave Macmillan, 2007); Hayden White, *Metahistory: The Historical Imagination in Nineteenth-Century Europe* (Baltimore: Johns Hopkins University Press, 1973). By "representations" I mean to

echo Stuart Hall's conception of "representation" as (1) an act of (re)construction rather than mere reflection; and (2) something conveying more than the surface or denotative meaning, but a deeper, mythlike connotation. Stuart Hall, "The Work of Representation," in Stuart Hall, ed., *Representation: Cultural Representations and Signifying Practices* (London: Sage, 1997), 13–74.

7. Michel Foucault, *Power/Knowledge: Selected Interviews and Other Writings, 1972–77*, edited by Colin Gordon (New York: Pantheon, 1980), 131–33. John Rennie Short discusses "regimes of representation" in "Urban Imagineers: Boosterism and the Representations of Cities," in Andrew E. G. Jonas and David Wilson, eds., *The Urban Growth Machine: Critical Perspectives Two Decades Later* (Albany: State University of New York Press, 1999), 38; and see Noam Chomsky, "Scholarship and Ideology: American Historians as 'Experts in Legitimation,'" *Social Scientist* 1, no. 7 (1973): 20–37.

8. Foucault, *Power/Knowledge*, 131.

9. Michel Foucault, "Nietzsche, Genealogy, History," in Donald Bouchard, ed., *Language, Counter-Memory, Practice: Select Essays and Interviews with Michel Foucault* (Ithaca, NY: Cornell University Press, 1980), 144–50.

10. Ralph Samuel, *Theatres of Memory: Past and Present in Contemporary Culture* (New York: Verso, 2012); and Henry Russo, *The Vichy Syndrome: History and Memory in France since 1944* (Cambridge, MA: Harvard University Press, 1991).

11. Samuel, *Theatres of Memory*, 16.

12. Historians Roy Rosenzweig and David Thelen conducted a survey of about 1,500 interviews asking people why and what they valued about history, as well as who/what they trusted most. Museums ranked first, with personal accounts from relatives second, conversation with a witness third, college history professors fourth, high school teachers fifth, nonfiction books sixth, and movies and television programs seventh. Roy Rosenzweig and David Thelen, *Presence of the Past: Popular Uses of History in American Life* (New York: Columbia University Press, 1998), 21.

13. For examples of analyzing a variety of mediums and representations of the past, see Aleida Assmann, "Canon and Archive," in Astrid Erll and Ansgar Nünning, eds., *Companion to Cultural Memory Studies* (New York: De Gruyter, 2010), 97–107; M. Christine Boyer, *The City of Collective Memory: Its Historical Imagery and Architectural Entertainments* (Cambridge, MA: MIT Press, 1994); Paul Connerton, *How Societies Remember* (New York: Cambridge University Press, 1989); David Glassberg, *Sense of History: The Places of the Past in American Life* (Amherst: University of Massachusetts Press, 2001); Jack Goody, "Memory in Oral and Literate Traditions," in Patricia Fara and Karalyn Patterson, eds., *Memory* (Cambridge: Cambridge University Press, 1998), 73–94; Marianne Hirsch, *Family Frames: Photography, Narrative and Postmemory* (Cambridge, MA: Harvard University Press, 1997); Eric Hobsbawm and Terence Ranger, eds., *The Invention of Tradition* (New York: Cambridge University Press, 1983); Iwona Irwin-Zarecka, *Frames of Remembrance: The Dynamics of Collective Memory* (New Brunswick, NJ: Transaction, 1994); Renate Lachman, "Cultural Memory and the Role of Literature," *European Review* 12, no. 2 (2004): 165–78; Denise Lawrence-Zuniga, *Protecting Suburban America: Gentrification, Advocacy and the Historic Imaginary* (New York: Bloomsbury, 2016); André Leroi-Gourhan, *Gesture and Speech* (Cambridge, MA: MIT Press 1996); George Lipsitz, *Time Passages: Collective Memory and American Popular Culture* (Minneapolis: University of Minnesota Press 1990); Martha Norkunas, *The Politics of Public Memory: Tourism, History, and Ethnicity in*

Monterey, California (Albany: State University of New York Press, 1993); Russo, *Vichy Syndrome*; Samuel, *Theatres of Memory*; Daniel J. Walkowitz and Lisa Maya Knauer, eds., *Memory and the Impact of Political Transformation in Public Space* (Durham, NC: Duke University Press, 2004); James V. Wertsch, *Voices of Collective Remembering* (New York: Cambridge University Press, 2002); James E. Young, *At Memory's Edge: After-Images of the Holocaust in Contemporary Art and Architecture* (New Haven, CT: Yale University Press, 2000); and Barbie Zelizer, *Covering the Body: The Kennedy Assassination, the Media, and the Shaping of Collective Memory* (Chicago: University of Chicago Press, 1992).

14. James E. Young, *The Texture of Memory: Holocaust Memorials and Meaning* (New Haven, CT: Yale University Press, 1993), xi. See also Eric Hobsbawm's notion of "invented tradition," Marianne Hirsch's "postmemory," and Paul Connerton's "structural amnesia." According to Hobsbawm, "'Invented tradition' is taken to mean a set of practices, normally governed by overtly or tacitly accepted rules and of a ritual or symbolic nature, which seek to inculcate certain values and norms of behavior by repetition, which automatically implies continuity with the past" (*Invention of Tradition*, 1). According to Hirsch, "Postmemory characterizes the experience of those who grow up dominated by narratives that preceded their birth, whose own belated stories are evacuated by the stories of the previous generation" (*Family Frames*, 22). Connerton presents structural amnesia as a type of forgetting that is, key, the result of remembering only what is available in print or other public recollections of the past ("Seven Types of Forgetting," *Memory Studies* 1, no. 1 [January 2008]: 64).

15. Jeffrey Olick, Vered Vinitzky-Seroussi, and Daniel Levy, "Introduction," in Jeffrey Olick, Vered Vinitzky-Seroussi, and Daniel Levy, eds., *Collective Memory Reader* (New York: Oxford University Press, 2011), 38; and Astrid Erll, "Cultural Memory Studies: An Introduction," in Erll and Nünning, *Companion to Cultural Memory Studies*, 2.

16. Cummins and Scheer, *Texan Identities*; Dan P. McAdams, *The Redemptive Self: Stories Americans Live By* (New York: Oxford University Press, 2006), 82.

17. For similar views, see Jan Assmann, *Moses the Egyptian: The Memory of Egypt in Western Monotheism* (Cambridge: Harvard University Press, 2009), 15; Connerton, *How Societies Remember*, 21; Ron Eyerman, "The Past in the Present: Culture and the Transmission of Memory," *Acta Sociologica* 47, no. 2 (2004): 159–69; David Lowenthal, "Identity, Heritage, and History," in John Gillis, ed., *Commemorations: The Politics of National Identity* (Princeton, NJ: Princeton University Press, 1994), 41–60; Alan Megill, *Historical Knowledge, Historical Error: A Contemporary Guide to Practice* (Chicago: University of Chicago Press, 2007), 42–62; and Jurgen Straub, *Narration, Identity and Historical Consciousness* (New York: Berghahn, 2006).

18. Alan Lambert, Laura Neese Scherer, Chad Rogers, and Larry Jacoby, "How Does Collective Memory Create a Sense of the Collective," in Paul Boyer and James V. Wertsch, eds., *Memory in Mind and Culture* (Cambridge, MA: Cambridge University Press, 2009), 194–95.

19. Idith Zertal, *Israel's Holocaust and the Politics of Nationhood* (Cambridge: Cambridge University Press, 2005), 2. See also Jan Assmann, "Collective Memory and Cultural Identity," *New German Critique* 65 (1995): 125–33; Diane Barthel, *Historic Preservation: Collective Memory and Historical Identity* (New Brunswick, NJ: Rutgers University Press, 1996); Dan Ben-Amos and Liliane Weissberg, eds., *Cultural*

Memory and the Construction of Identity (Detroit, MI: Wayne State University Press, 1999); Glassberg, *Sense of History*; Michael Kammen, *Mystic Chords of Memory: The Transformation of Tradition in American Culture* (New York: Knopf 1991); Lowenthal, "Identity, Heritage, and History"; and Walkowitz and Knauer, *Memory and the Impact*.

20. Benedict Anderson, *Imagined Communities: Reflections on the Origin and Spread of Nationalism* (New York: Verso, 1983), 1–5.

21. For just a few examples, see John W. Store and Mary L. Scheer, eds., *Twentieth-Century Texas: A Social and Cultural History* (Denton: University of North Texas Press, 2008); Donald Willett and Stephen J. Curley, *Invisible Texans: Women and Minorities in Texas History* (Boston: McGraw-Hill, 2004); Félix D. Almaráz, Patricia Thompson, and Carmen Reyes-Johnson, *Eyewitnesses to Texas History* (Dubuque, IA: Kendall/Hunt, 2008); Ty Cashion and Jesús F. de la Teja, *The Human Tradition in Texas* (Lanham, MD: Rowman and Littlefield, 2001); Bruce A. Glasrud and James C. Maroney, *Texas Labor History* (College Station: Texas A&M University Press, 2013); Cullen and Wilkison, *Texas Left*; Maggie Rivas-Rodriguez, *Texas Mexican Americans and Postwar Civil Rights* (Austin: University of Texas Press, 2015); and Sandra M. Mayo and Elvin Holt, *Stages of Struggle and Celebration: A Production History of Black Theatre in Texas* (Austin: University of Texas Press, 2016). For a review of more works, see essays in Glasrud, Cummins, and Wintz, *Discovering Texas History*.

22. For a review of urban history in Texas, see Richard B. Wright, "Texas Urban History," in Glasrud, Cummins, and Wintz, *Discovering Texas History*, 134–49 (quote 135); and Cheryl Caldwell Ferguson, *Highland Park and River Oaks: The Origins of Garden Suburban Community Planning in Texas* (Austin: University of Texas Press, 2016).

23. Scarbrough, *Road, River*; Cunningham, *Cowboy Conservatism*; and Bill Bishop, *The Big Sort: Why the Clustering of Like-Minded America Is Tearing Us Apart* (Boston: Houghton Mifflin, 2008).

24. The literature on spatial theory is vast, perhaps too vast. But for works that are useful in discussing space as something more than a material landscape (sometimes called material space), including how space can be represented, even imagined (often called imagined space, storied space), as well as lived and experienced (sometimes called lived space or practiced space), see Michel de Certeau, *The Practice of Everyday Life* (Berkeley: University of California Press, 1984), esp. part 3; Steven Feld and Keith Basso, eds., *Senses of Place* (Santa Fe, NM: School of American Research Press, 1996); Dolores Hayden, *The Power of Place: Urban Landscapes as Public History* (Cambridge, MA: MIT Press, 1996); Henri Lefebvre, *The Production of Space* (1974; repr., Oxford, UK: Blackwell, 1991); Gyan Prakash and Kevin Kruse, eds., *The Spaces of the Modern City: Imaginaries, Politics, and Everyday Life* (Princeton, NJ: Princeton University Press, 2008); Edward Soja, *Thirdspace: Journeys to Los Angeles and Other Real-and-Imagined Places* (Cambridge: Blackwell, 1996); and Yi-Fu Tuan, *Space and Place: The Perspective of Experience* (Minneapolis: University of Minnesota, 1977).

25. John Archer, "The Place We Love to Hate: The Critics Confront Suburbia, 1920–1960," in Klaus Stierstorfer, ed., *Constructions of Home: Interdisciplinary Studies in Architecture, Law, and Literature* (New York: AMS Press, 2010), 45–82. For more reviews of critics, and some of their work, see Becky M. Nicolaides and Andrew Wiese, eds., *The Suburb Reader* (New York: Routledge, 2006). See also James Howard Kunstler, *The Geography of Nowhere: The Rise and Decline of America's Man-Made*

Landscape (New York: Simon and Schuster, 1993), 273 (quote); Betty Friedan, *The Feminine Mystique* (New York: Norton, 1963); and William H. Whyte, *The Organization Man* (New York: Simon and Schuster, 1956).

26. For suburbia as cultural practice, see Robert Fishman, *Bourgeois Utopias: The Rise and Fall of Suburbia* (New York: Basic Books, 1987), 8–9; Dolores Hayden, *Building Suburbia: Green Fields and Urban Growth, 1820–2000* (New York: Vintage Books, 2003), 3, 8; Mary Corbin Sies, "Moving beyond Scholarly Orthodoxies in North American Suburban History," *Journal of Urban History* 27, no. 3 (March 2001): 355–61; and Mary Corbin Sies, "The City Transformed: Nature, Technology, and the Suburban Ideal, 1877–1917," *Journal of Urban History* 14, no. 1 (November 1987): 81–111.

27. For more on popular cultural representations of suburbs, see Robert Beuka, *Suburbia Nation: Reading Landscape in Twentieth-Century American Fiction and Film* (New York: Palgrave, 2004); David Forrest, Graeme Harper, and Jonathan Rayner, eds., *Filmurbia: Screening the Suburbs* (London: Palgrave Macmillan, 2017); Rupa Huq, *Making Sense of Suburbia through Popular Culture* (New York: Bloomsbury, 2013); Catherine Jurca, *White Diaspora: The Suburbs and the Twentieth-Century American Novel* (Princeton, NJ: Princeton University Press, 2001); and Lynn Spigel, *Make Room for TV: Television and the Family Ideal in Postwar America* (Chicago: University of Chicago Press, 1992).

28. Irwin Silverman and Jean Choi, "Locating Places," in David Buss, ed., *The Handbook of Evolutionary Psychology* (Hoboken: John Wiley and Sons, 2005), 189–94.

29. For more on the rural ideal, see David Allmendinger Jr., *Ruffin: Family and Reform in the Old South* (New York: Oxford University Press, 1990); John Archer, "Country and City in the American Romantic Suburb," *Journal of the Society of Architectural Historians* 42, no. 2 (May 1983): 139–56; Daniel Boorstin, *The Lost World of Thomas Jefferson* (New York: H. Holt, 1948); Joyce E. Chaplin, *An Anxious Revolt: Agricultural Innovation and Modernity in the Lower South, 1730–1815* (Chapel Hill: University of North Carolina Press, 1993); William Cronon, "The Trouble with Wilderness," *New York Times Magazine*, August 14, 1995, 46–47; David Danbom, *Born in the Country: A History of Rural America*, 2nd ed. (Baltimore: Johns Hopkins University Press, 2006), 65–69; Clarence Danhof, *Change in Agriculture: The Northern United States, 1820–1870* (Cambridge, MA: Harvard University Press, 1969); Fishman, *Bourgeois Utopias*, 53–54, 127; R. Douglas Hurt, *American Agriculture: A Brief History*, rev. ed. (West Lafayette, IN: Purdue University Press, 2002), 72–77; Kenneth Jackson, *Crabgrass Frontier: The Suburbanization of the United States* (New York: Oxford University Press, 1985), 57; Leo Marx, *The Machine in the Garden: Technology and the Pastoral Ideal in America* (New York: Oxford University Press, 1964); Morrill Peterson, *The Jefferson Image in the American Mind* (New York: Oxford University Press, 1960); and Raymond Williams, *The Country and the City* (New York: Oxford University Press, 1973).

30. See Danbom, *Born in the Country*, 65–66.

31. Thomas Jefferson, quoted in Charles Glaab and A. Theodore Brown, *A History of Urban America* (New York: Macmillan, 1967), 55.

32. For more on the meaning of the cowboy, particularly in Texas, see John Bainbridge, *The Super-Americans: A Picture of Life in the United States, as Brought into Focus, Bigger Than Life, in the Land of the Millionaires—Texas* (Garden City, NY: Doubleday, 1961); Dee Brown, *The American West* (New York: Scribner, 1994); Paul

Carlson, ed., *The Cowboy Way: An Exploration of History and Culture* (Lubbock: Texas Tech University Press, 2000); Jay Paul Childers, "Cowboy Citizenship: The Rhetoric of Civic Identity among Young Americans, 1965–2005," Ph.D. dissertation, University of Texas, 2006; David Dary, *Cowboy Culture: A Saga of Five Centuries* (New York: Knopf, 1981); Eric Hobsbawm, "The American Cowboy: An International Myth?," in *Fractured Times: Culture and Society in the Twentieth Century* (London: Abacus, 2014); Judith Kleinfeld and Andrew Kleinfeld, "Cowboy Nation and American Character," *Society Journal* 41, no. 2 (March/April 2004): 43–50; Paul Reddin, *Wild West Shows* (Urbana: University of Illinois Press, 1999); and Richard Slotkin, *Gunfighter Nation: The Myth of the Frontier in Twentieth-Century America* (Norman: University of Oklahoma Press, 1998).

33. Andrew C. Baker, "From Rural South to Metropolitan Sunbelt: Creating a Cowboy Identity in the Shadow of Houston," *Southwestern Historical Quarterly* 118, no. 1 (July 2014): 1–22; Kleinfeld and Kleinfeld, "Cowboy Nation," 43–50; and Slotkin, *Gunfighter Nation.*

34. Randolph Campbell, *An Empire for Slavery: The Peculiar Institution in Texas, 1821–1865* (Baton Rouge: Louisiana University Press, 1989), 1; Baker, "From Rural South"; and Cantrell and Turner, *Lone Star Pasts.*

35. Baker, "From Rural South," 3.

36. For more on Reagan and the cowboy image, see Laura R. Barraclough, *Making the San Fernando Valley: Rural Landscapes, Urban Development, and White Privilege* (Athens: University of Georgia Press, 2011), 110–12.

37. See Todd Gardner, "The Slow Wave: The Changing Residential Status of Cities and Suburbs in the United States, 1850–1949," *Journal of Urban History* 27, no. 3 (March 2001): 296–99; Hayden, *Building Suburbia*, 249; Nicolaides and Wiese, *Suburb Reader*, 2; J. John Palen, *The Suburbs* (New York: McGraw-Hill, 1995), 11–12; and U.S. Bureau of Census, Census of Population and Housing: 2000, vol. 1, *Summary of Population and Housing Characteristics*, pt. 1 (Washington, DC: Government Printing Office, 2002), app. A-16. See also Myron Orfield, *American Metropolitics: The New Suburban Reality* (Washington, DC: Brookings Institution Press, 2002); Peter Dreir, John Mollenkopf, and Todd Swanstrom, *Place Matters: Metropolitics for the Twenty-First Century* (Topeka: University of Kansas Press, 2002); and Bruce Katz and Robert E. Lang, eds., *Redefining Urban and Suburban America: Evidence from Census 2000* (Washington, DC: Brookings, 2003). On shopping and job location, see Palen, *Suburbs*, 5; William Sharpe and Leonard Wallock, "Bold New City or Built-Up 'Burb? Redefining Contemporary Suburbia," *American Quarterly* 46, no. 1 (March 1994): 2; and Richard Harris and Robert Lewis, "The Geography of North American Cities and Suburbs, 1900–1950: A New Synthesis," *Journal of Urban History* 27, no. 3 (March 2001): 263. Actually, the trend was identified in the mid-1980s by Mark Baldassare, *Trouble in Paradise: The Suburban Transformation in America* (New York: Columbia University Press, 1986), 7; and William K. Stevens, "Beyond the Mall: Suburbs Evolving into 'Outer City,'" *New York Times*, November 8, 1987, E5.

38. "*Megaregions*," *America2050*, a project of the Regional Plan Association, www.america2050.org/megaregions.html. See also Robert E. Lang and Dawn Dhavale, "Beyond Megalopolis: Exploring America's New 'Megapolitan' Geography," Metropolitan Institute at Virginia Tech Census Report Series (July 2005), https://digitalscholarship.unlv.edu/brookings_pubs/38.

39. Harrison Potter, "Cities or Suburbs: Which Area Is Seeing a Population Boom," *Nation Swell*, July 3, 2014, http://nationswell.com/cities-suburbs-population -growth-urban-areas.

40. Jed Kolko, "The 10 Healthiest U.S. Housing Markets Going into 2013," *Atlantic*, Dec 13, 2012, www.citylab.com/housing/2012/12/10-healthiest-housing-markets -2013/4154; U.S. Census Bureau, Cumulative Estimates of Resident Population Change for Incorporated Places Over 50,000, Ranked by Percent Change: April 1, 2010 to July 1, 2012—United States–Places of 50,000+ Population, May 2013, http:// factfinder.census.gov/faces/tableservices/jsf/pages/productview.xhtml?src=bkmk; William H. Frey, "Texas Gains, Suburbs Lose in 2010 Census Preview," June 25, 2010, *Brookings*, www.brookings.edu/research/opinions/2010/06/25-population -frey; Young, "Census: Suburbs"; Renfro, "Texas Dominates List"; Kriel and Begley, "Census Data Show [Houston] City"; Kotkin, "America's Fastest-Growing Counties"; and Cox, "Texas Suburbs Lead Population Growth."

41. Cox, "Texas Suburbs Lead Population Growth"; Josh Sanburn, "U.S. Cities Are Slowing but Suburbs Are Growing," *Time*, May 22, 2014, http://time.com/107808 /census-suburbs-grow-city-growth-slows; and Nate Berg, "Urban vs. Suburban Growth in U.S. Metros," *Atlantic*, June 29, 2012, www.citylab.com/housing/2012/06/urban -or-suburban-growth-us-metros/2419.

42. Greg Dickinson, *Suburban Dreams: Imagining and Building the Good Life* (Tuscaloosa: University of Alabama Press. 2015), 154–81. For an example in Houston, see Rives J. Taylor, "Lifestyle Centers Coming (Back) to Houston," *Cite* 65 (Winter 2005): 26–31.

43. Jill H. Wilson and Nicole Prchal Svajlenka, "Immigrants Continue to Disperse, with Fastest Growth in the Suburbs," Immigration Facts Series, *Brookings*, October 29, 2014, www.brookings.edu/research/papers/2014/10/29-immigrants -disperse-suburbs-wilson-svajlenka.

44. John Sullivan, "African Americans Moving South—and to the Suburbs," *Autumn Awakening* 18, no. 2 (2011), http://reimaginerpe.org/18–2/sullivan; and Dan Solomon, "Why It Matters That Austin's Black Population Is Being Pushed to the Suburbs," *Texas Monthly*, January 28, 2015, www.texasmonthly.com/daily-post/why -it-matters-austins-black-population-being-pushed-suburbs.

45. Josh Sanburn, "The Rise of Suburban Poverty in America," *Time*, July 31, 2014, http://time.com/3060122/poverty-america-suburbs-brookings; and Hope Yen, "'Bright Flight' Changes the Face of Cities, Suburbs," *NBCNews.com*, May 9, 2010, www.nbcnews.com/id/37041770/ns/us_news-census_2010/t/bright-flight-changes -face-cities-suburbs.

46. Juan Castillo and Melissa B. Taboada, "Poverty Takes Root in Austin's Suburbs," *Austin American-Statesman*, May 19, 2013, www.mystatesman.com/news/news /local/poverty-takes-root-in-austins-suburbs/nXwt2.

47. Sullivan, "African Americans Moving South"; Solomon, "Why It Matters"; and Sanburn, "Rise of Suburban Poverty."

48. Buenger and León, *Beyond Texas through Time*, xx.

49. Hubert Howe Bancroft authored, or sponsored, thirty-nine volumes on regional/state history from 1874 to 1890. The Lewis Historical Publishing Company, which no longer exists, published an untold number of local, regional, and state histories in the early to mid-twentieth century, ranging from Baltimore to Colorado to San

Francisco, which also featured short place biographies of suburban areas surrounding major metropolises. Not mentioned in the text are the innumerable state and regionally based publishers that also produced legions of county, state, and regional histories that often highlighted suburban communities (even if not always labeled as such). Likewise, city and county directories, particularly those published by R. L. Polk and Co. starting in 1872, contain innumerable short histories, or place biographies, of suburban communities across the United States (usually written by local boosters who too often simply cut and pasted the text found in larger histories written by Bancroft et al).

50. Among the many titles for Walter Prescott Webb, see *The Great Plains* (Boston: Ginn, 1931); *The Great Frontier* (Boston: Houghton Mifflin, 1952); and *The Handbook of Texas*, 3 vols. (Austin: Texas State Historical Association, 1952–76). For a good review of sectionalism, also often labeled "regionalism," particularly in lieu of Frederick Jackson Turner's influence, see T. R. C. Hutton, "Beating a Dead Horse: The Continuing Presence of Frederick Jackson Turner in Environmental and Western History," *International Social Science Review* 77, nos. 1–2 (2002): 47–57; and Richard Jensen, "On Modernizing Frederick Jackson Turner: The Historiography of Regionalism," *Western Historical Quarterly* 11, no. 3 (Fall 1980): 307–22.

51. Herbert J. Gans, *The Levittowners: Ways of Life and Politics in a New Suburban Community* (New York: Pantheon Books, 1967); Whyte, *Organization Man*; and Lewis Mumford, "Suburbia—and Beyond," in *The City in History: Its Origins, Its Transformations, and Its Prospects* (New York: Harcourt, Brace and World, 1961), 483–503. For a review of critics of suburbia in the early twentieth century through the post–World War II era, see Archer, "Place We Love to Hate." For other critics spanning the centuries, see Nicolaides and Wiese, *Suburb Reader.*

52. Stephan Thernstrom, *Poverty and Progress: Social Mobility in a Nineteenth-Century City* (Cambridge, MA: Harvard University Press, 1964). The so-called old urban history is summarized well by Charles N. Glaab, "The Historian and the American City: A Bibliographic Survey," in Philip M. Hauser and Leo F. Schnore, eds., *The Study of Urbanization* (New York: John Wiley, 1965), 53–80. For early examples of the new urban history, besides Thernstrom's *Poverty and Progress*, see Sam Bass Warner Jr., *The Urban Wilderness: A History of the American City* (New York: Harper and Row, 1972); Stephan Thernstrom, *The Other Bostonians: Poverty and Progress in the American Metropolis, 1880–1970* (Cambridge, MA: Harvard University Press, 1973); Eric H. Monkkonen, *The Dangerous Class: Crime and Poverty in Columbus Ohio, 1860–1865* (Cambridge, MA: Harvard University Press, 1975); Michael B. Katz, *The People of Hamilton, Canada West* (Cambridge, MA: Harvard University Press, 1975); Kathleen N. Conzen, *Immigrant Milwaukee, 1836–1860: Accommodation and Community in a Frontier City* (Cambridge, MA: Harvard University Press, 1976); and Stuart M. Blumin, *The Urban Threshold: Growth and Change in a Nineteenth-Century American Community* (Chicago: University of Chicago Press, 1976). For reviews of new urban history, see Theodore Hershberg's classic essay on the need to see the city as both place and process, "The New Urban History: Toward an Interdisciplinary History of the City," *Journal of Urban History* 5, no. 1 (November 1978): 3–40; and Thernstrom's review, "The New Urban History," in Charles F. Delzell, ed., *the Future of History: Essays in the Vanderbilt University Centennial Symposium* (Nashville, TN: Vanderbilt University Press, 1977), 43–52.

53. Jon C. Teaford, *City and Suburb: The Political Fragmentation of Metropolitan America, 1850–1970* (Baltimore: Johns Hopkins University Press, 1979); and Sam Bass

Warner Jr., *Streetcar Suburbs: The Process of Growth in Boston, 1870–1900* (Cambridge, MA: Harvard University Press, 1962). For examples of urban historians (or at least urban histories) producing metropolitan histories, see Carl Abbott, *Metropolitan Frontier: Cities in the Modern American West* (Tuscon: University of Arizona Press, 1993); William Cronon, *Nature's Metropolis: Chicago and the Great West* (New York: Norton, 1991); and Kenneth Fox, *Metropolitan America: Urban Life and Urban Policy in the United States, 1940–1980* (Jackson: University Press of Mississippi, 1986). For earlier historians and authors who focused (briefly) on suburbs and who also took a metropolitan perspective, see Glenn Dumke, *The Boom of the Eighties in Southern California* (San Marino, CA: Huntington Library, 1944); Fishman, *Bourgeois Utopias*; Carey McWilliams, *Southern California: An Island on the Land* (New York: Duell, Sloan and Pearce, 1946), esp. chap. 11; Teaford, *City and Suburb*; and Warner, *Streetcar Suburbs*, esp. 153–56. Many recent suburban historians are also taking a broader metropolitan view to pay more attention to the place of the suburbs in relationship with central cities, competing suburbs, and their region as a whole. See Eric Avila, *Popular Culture in the Age of White Flight: Fear and Fantasy in Suburban Los Angeles* (Berkeley: University of California Press, 2006); Hayden, *Building Suburbia*, esp. 4–10; Greg Hise, *Magnetic Los Angeles: Planning the Twentieth-Century Metropolis* (Baltimore: Johns Hopkins University Press, 1997); Matthew Lassiter, "The New Suburban History II: Political Culture and Metropolitan Space," *Journal of Planning History* 4, no. 1 (February 2005): 75–88; Matthew Lassiter, *The Silent Majority: Suburban Politics in the Sunbelt South* (Princeton, NJ: Princeton University Press, 2006); Robert Lewis, ed., *Manufacturing Suburbs: Building Work and Home on the Metropolitan Fringe* (Philadelphia: Temple University Press, 2004); Paul H. Mattingly, *Suburban Landscapes: Culture and Politics in a New York Metropolitan Community* (Baltimore: Johns Hopkins University Press, 2001); Becky Nicolaides, *My Blue Heaven: Life and Politics in the Working-Class Suburbs of Los Angeles, 1920–1965* (Chicago: University of Chicago Press, 2002), esp. chap. 3; Nicolaides and Wiese, *Suburb Reader*, 8; Margaret Pugh O' Mara, "Suburbia Reconsidered: Race, Politics, and Prosperity in the Twentieth Century," *Journal of Social History* 39, no. 1 (Fall 2005): 229–44; Robert Self, *American Babylon: Race and the Struggle for Postwar Oakland* (Princeton, NJ: Princeton University Press, 2003); Amanda Seligman, "The New Suburban History," *Journal of Planning History* 3, no. 4 (November 2004): 312–33; and Sies, "City Transformed."

54. A large body of works have focused on African Americans, minorities, industrial deconcentration, and working suburbs. These works not only challenge the stereotype that suburbs are for the middle class but also demonstrate a trend in examining nonwhite suburbs that the new suburban historians claim an earlier cohort reified. See Margaret Crawford, *Building the Workingman's Paradise: The Design of American Company Towns* (New York: Verso, 1995); Timothy Fong, *The First Suburban Chinatown: The Remaking of Monterey Park, California* (Philadelphia: Temple University Press, 1994); Matt Garcia, *A World of Its Own: Race, Labor, and Citrus in the Making of Greater Los Angeles, 1900–1970* (Chapel Hill: University of North Carolina Press, 2001); Richard Harris, *Unplanned Suburbs: Toronto's American Tragedy, 1900 to 1950* (Baltimore: Johns Hopkins University Press, 1996); Richard Harris and Robert Lewis, "The Geography of North American Cities and Suburbs, 1900–1950: A New Synthesis," *Journal of Urban History* 27, no. 3 (March 2001): 262–93; Hayden, *Building Suburbia* , 97–127; Hise, *Magnetic Los Angeles*; Wei Li, "Building Ethnoburbia: The

Emergence and Manifestation of the Chinese Ethnoburb in Los Angeles' San Gabriel Valley," *Journal of Asian American Studies* 2, no. 1 (February 1999): 1–29; Lewis, *Manufacturing Suburbs*; Nicolaides, *My Blue Heaven*; Mary Pattillo-McCoy, *Black Pickett Fences: Privilege and Peril among the Black Middle Class* (Chicago: University of Chicago Press, 1999); Thomas Sugrue, *The Origins of the Urban Crisis: Race and Inequality in Postwar Detroit* (Princeton, NJ: Princeton University Press, 1996); Alexander von Hoffman, *Local Attachments: The Making of an American Urban Neighborhood, 1850–1920* (Baltimore: Johns Hopkins University Press, 1994); Andrew Wiese, *Places of Their Own: African American Suburbanization in the Twentieth Century* (Chicago: University of Chicago Press, 2004); and William Wilson, *Hamilton Park: A Planned Black Community in Dallas* (Baltimore: Johns Hopkins University Press, 1998).

55. Barraclough, *Making the San Fernando Valley*; Matthew Gordon Lasner, *High Life Condo Living in the Suburban Century* (New Haven, CT: Yale University Press, 2012); Christopher Sellers, *Crabgrass Crucible: Suburban Nature and the Rise of Environmentalism in Twentieth-Century America* (Durham: University of North Carolina Press, 2012); Wendy Cheng, *The Changs Next Door to the Díazes: Remapping Race in Suburban California* (Minneapolis: University of Minnesota Press, 2013); Lily Geismer, *Don't Blame Us: Suburban Liberals and the Transformation of the Democratic Party* (Princeton, NJ: Princeton University Press, 2014); Paul J. P. Sandul, *California Dreaming: Boosterism, Memory, and Rural Suburbs in the Golden State* (Morgantown: West Virginia University Press, 2014); John Archer, Paul J. P. Sandul, and Katherine Solomonson, eds., *Making Suburbia: New Histories of Everyday America* (Minneapolis: University of Minnesota Press, 2015); James A. Jacobs, *Detached America: Building Houses in Postwar Suburbia* (Charlottesville: University of Virginia Press, 2015); Vicki Howard, *From Main Street to Mall: The Rise and Fall of the American Department Store* (Philadelphia: University of Pennsylvania Press, 2015); Dickinson, *Suburban Dreams*; Lisa Uddin, *Zoo Renewal: White Flight and the Animal Ghetto* (Minneapolis: University of Minnesota Press, 2015); Lawrence-Zuniga, *Protecting Suburban America*; and Willow S. Lung-Amam, *Trespassers? Asian Americans and the Battle for Suburbia* (Berkeley: University of California Press, 2017).

56. Archer, Sandul, and Solomonson, "Introduction," in *Making Suburbia*, vii–xxv.

57. De Certeau, *Practice of Everyday Life*, xix, 100, 117.

58. This echoes Claude Lévi-Strauss's conception of a *bricoleur* who creates and improvises (even alters, distorts, adds, deletes, reorders, transforms, etc.) by appropriating existing materials and resources that are readily available and known. *The Savage Mind* (Chicago: University of Chicago Press, 1966), 16–33.

59. Archer, Sandul, and Solomonson, "Introduction," in *Making Suburbia*, vii–xxv; and Bruno Latour, *Reassembling the Social: An Introduction to Actor-Network Theory* (New York: Oxford University Press, 2005), 12.

{ 2 }

PLANNING THE SUBURBAN CITY IN THE DALLAS—FORT WORTH METROPLEX

ROBERT B. FAIRBANKS

Although historians have recently introduced us to the complexity of suburbia after World War II, surprisingly little has been done on the role of planning in these emerging communities.[1] Many suburbs had been small agricultural communities on the outskirts of Dallas and Fort Worth before the war but changed drastically after V-J Day as they started their path to becoming suburban cities. Little attention has been given to the planning associated with the communities that become boom suburbs in the 1950s and 1960s. What eventually becomes known as the Dallas–Fort Worth Metroplex provides an excellent setting to explore the extent and impact of planning in those communities that would become cities in their own right.

According to the estimates of the North Central Texas Council of Governments, there are eleven suburban cities in the Dallas–Fort Worth Metroplex, each with a population of over 100,000. Four of them—Arlington, Plano, Garland, and Irving—had more than 200,000 residents.[2] Moreover, according to the most recent estimates, the Dallas–Fort Worth–Arlington metropolitan area is the fourth most populous in the nation after the New York, Los Angeles, and Chicago metropolitan statistical areas, and Arlington is the fiftieth-largest city in the nation, larger than Cincinnati or Buffalo.

Although a growing scholarly literature has recognized the emergence of these new types of urban/suburban places,[3] it appears that most scholars would agree with Robert E. Lang and Jennifer B. LeFurgy, who subtitled their book "The Rise of America's Accidental Cities."[4] Nevertheless, a close look at many of these places in North Texas suggests that leadership and the ability to take advantage of their location within a dynamic metropolitan region meant that their emergence was more than simply accidental. It required strong leadership and a vision of the type of cities these new places wanted to be. Toward that goal, all four of the communities studied here committed to planning at an early stage, and those plans suggested a vision that went beyond becoming simply a bedroom suburb or an accidental city. Promoters of communities like Arlington, Irving, Garland, and Grand Prairie had bigger dreams and created a new type of city—a suburban city—by seeing their positioning within the metropolitan region as an advantage and the metropolitan area itself as a type of hinterland that could be exploited for growth and development.

Just as nineteenth-century cities emerged in part because of strong leadership, transportation developments, and a rural hinterland that allowed growth, so did suburban cities of the second half of the twentieth century. Such communities did not just passively sit back and wait for growth to happen but rather pursued it, just as nineteenth-century boosters had done for their communities. For instance, North Texas suburban communities in the 1950s and 1960s undertook aggressive annexation campaigns under the state's liberal annexation laws for home rule cities, increasing the size of their communities and finding room for industry that would not interfere with high-quality residential development. They also took advantage of the state law that allowed them to reserve land for future development by initiating but not completing annexation proceedings ("first reading" of intent to annex). Furthermore, they embraced planning often in the early 1950s, and this resulted in master plans that focused on ordered commercial and industrial development as well as planned residential growth. Those places that had deteriorating black neighborhoods also strategized how to improve those areas. In addition, these suburban towns, often through the efforts of their chambers of commerce, lobbied for better highways, more water, and effective sewage disposal systems so they could attract industry and commerce to their communities.

Civic leaders in these small towns understood that the post–World War II setting offered an opportunity to develop their communities into midsize cities that would thrive and prosper in the metropolitan setting. Their location provided them unique opportunities that other small but ambitious communities did not have. As a result, town boosters adopted a mentality that embraced strategies that would guarantee quality-of-life issues along with significant growth, mixed land use, and a healthy tax base.

Although scholars have examined planned suburban communities such as Forest Park, Ohio, Irvine, California, and Park Forest, Illinois,[5] little attention has been given to planning in the small communities near big cities as they started to experience explosive growth after World War II. This neglect is unfortunate not only for the history of suburbia but for the history of planning in the United States, since much of the planning profession's growth at this time came about through a growing interest in planning in small communities encouraged by the 701 Program created by the Housing Act of 1954. This chapter starts to address that neglect by examining the role of planning in the emergence of suburban cities in the Dallas–Fort Worth metropolitan complex after World War II.

It might startle some to hear that in conservative Texas twenty suburban communities in the Dallas area engaged planners and developed plans between 1950 and 1963.[6] Not all of these communities had ambitions to become large suburban cities, but they all saw planning as a way to avoid mistakes that had been made by older cities. Suburbs such as Grand Prairie, Arlington, Irving, and Garland created planning commissions as soon as they passed the 5,000-population figure, the minimum required to become a home rule city. It was probably no accident that these four communities took the lead in becoming suburban cities, since they all experienced rapid growth during or immediately after the war.

Grand Prairie, located ten miles west of Dallas and described in 1940 as a "sleepy country town" of 1,595 that served local cotton farmers, experienced tremendous growth thanks to World War II and this nation's defense program. In 1940, federal officials located a giant North American Aviation airplane factory just east of the city's boundaries. During the peak of the war it employed more than 35,000 workers and promoted the rapid growth of the community.[7] According to B. A. Stufflebeme, president of the First National Bank of Grand Prairie, 95 percent of the homes built in Grand Prairie were the result of the airplane factory.[8] Adjacent to the plant sat a naval air station (Hensley Field), drawing still additional population and development to the area. As a result, by 1950 the town's population had accelerated to 14,594, an 815 percent increase and the greatest percentage of growth in the state. Nearby Arlington, just west of Grand Prairie and nineteen miles from Dallas, also benefited from the need for additional defense workers' housing. Its population rose from 4,240 to 7,692 as many located in Arlington to work at the airplane factories in Grand Prairie and Fort Worth. Garland, located fourteen miles northeast of downtown

Dallas and home to significant war industry, grew 373 percent from 2,233 to 10,571, making it the third-fastest-growing community in Texas during the war decade. Irving, another rural town located northeast of Grand Prairie and just west of Dallas, also more than doubled its population during the 1940s to 2,621, thanks in part to nearby war plants.[9]

Even though some of the defense plants closed after the war, industrial decentralization continued and brought numerous migrants to the area; this along with a spike in babies by returning servicemen created a dynamic setting in North Texas in the 1950s.[10] Indeed, all four of the communities studied here experienced rapid growth that decade and turned to city planning not just to cope with their increasing populations but to take advantage of the opportunities that such growth throughout the metropolitan area offered. The communities established planning commissions and engaged outside experts to help plan even before Congress created the Section 701 Program of the Housing Act of 1954. Arlington, possibly the most ambitious of the four towns, completed the area's first suburban master plan in 1952 after employing Robert W. Caldwell, a planner associated with the Fort Worth engineering firm of Freese and Nichols. The city paid $6,750 to Caldwell, formerly a planner for the Tennessee Valley Authority and more recently a planner of other midsize Texas towns, to prepare the 125-page document.[11]

Twenty-six-year-old mayor Tom Vandergriff led the movement for planning and was the community's prime booster. After the war, as both Fort Worth and Dallas experienced unprecedented growth, some Arlington civic leaders realized the community's unique position to benefit from the region's development since it sat midway between the two cites on U.S. 80, the main highway connecting Dallas and Fort Worth. No one saw the potential for growth better than Vandergriff, son of a respected car dealer and recent graduate of the University of Southern California. When he came home to help run his father's General Motors automobile dealership, Vandergriff, then twenty-three, became president of the town's chamber of commerce in 1949 and a vocal advocate of making Arlington more than merely a small town or bedroom suburb.[12] While living in Los Angeles, Vandergriff had witnessed the transforming power of the automobile in Orange County and understood that Arlington would likewise benefit from the coming sprawl.[13] He and fellow chamber members wanted to guide the community so it would become a self-sufficient city within the expanding metropolitan area. This meant attracting business and industry as well as residents to the small agricultural community. While still chamber president, Vandergriff started negotiating with General Motors to build an automobile plant in Arlington using the community's location as a key selling point. His desire to attract the car plant led him to run for mayor in 1952, and his growth platform helped him defeat the incumbent.[14] As mayor he immediately turned to planning as a key to the city's ordered growth.

The local Arlington newspaper treated the completion of the city's first plan in 1952, initiated in October 1951, as a major event in the community's history. Numerous editorials and front-page stories highlighted the planning effort. Reports both stressed how planning would allow the city to avoid some of the mess that other North Texas suburbs had faced but also viewed the plan as a booster tool, emphasizing how it marked the community as "progressive" and promised future businesses an orderly and efficient environment. The discussion predicted that planning would promote residential as well as economic growth and development besides improving citizens' quality of life. The planning document concluded that "through the use of the long-range Plan and its methods of controls" Arlington would be "patterned into the kind of City that citizens want it to be."[15]

As part of the planning process, Caldwell also helped produce the suburb's first comprehensive zoning ordinance. It included ten classifications and a zoning map that roughly followed Ernest Burgess's concentric zone theory of city development. It zoned the downtown area for business and commerce, designated areas near the core for multiunit housing, and zoned much of the rest of the outlying land for single-family homes.[16]

At the time of the plan, the suburb had a little over two hundred blacks living within the municipal corporation just northwest of the downtown area called "The Hill."[17] Anticipating the continued growth of the black population in Arlington, Caldwell suggested developing a new residential subdivision for African Americans in southeast Arlington near the Grand Prairie African American neighborhood. According to the planner, "Such a development would serve to attract the settlement of these people in a more suitable area than they are now located.[18] Light industry, such as an area proposed for the General Motors plant, was located near the Texas and Pacific Railroad tracks farther away from the urban core. Despite controversies in other Texas cities over zoning (notably Houston), the city's zoning ordinance passed rather easily with a minimum amount of protest.[19]

Meanwhile, officials presented the completed city plan to the public at a ceremony in the auditorium of the First National Bank building on September 24, 1952. Nothing in that plan stood out as particularly path breaking or distinguishing, but it did mark the first time the town had a document to guide its future physical growth. After an introduction and a chapter on the city's population and economic base, the plan examined the city's land-use patterns and offered a future land-use plan and zoning ordinance. In addition, it prescribed subdivision control and included chapters on plans for major streets and highways, automobile parking, airports, public school building location, parks, playgrounds, and public building sites as well as public utilities. The final chapter on the administration of the plan included strategies for citizen support and offered a long-term financial program. The longest chapter focused on street development to create through streets with uniform widths, resolving problems that had surfaced in the late 1940s before officials had approved adequate subdivision control. It called for the improvement of nine major north-south thoroughfares along with eight east-west arteries.[20]

Grand Prairie, a community of five square miles, also turned to planning in part to cope with the explosive growth that the war brought the city and to order the continuing postwar growth. Shortly after Dallas suddenly annexed the site of the North American Aviation factory in 1947, stealing from Grand Prairie a valuable site for industrial use and tax revenue, Grand Prairie became a home rule city so that it could annex nearby land.[21] After twelve-term mayor G. H. Turner announced his retirement in 1949, E. Carlyle Smith ran on a platform calling for city planning and won the election. Shortly after that he appointed a fourteen-citizen committee to help create a master plan, and the city eventually employed William Llewellyn Powell of Powell and Powell Engineers, Dallas, as the planning consultant.

That action resulted in the city's first plan, published in 1954. Even before that, officials approved the city's first city zoning ordinance in March 1950 and three years later developed the city's first major street plan.[22] In 1953, voters selected a younger group of civic leaders to commission government, known as the "Whiz Kids" because of their age, and this group successfully pushed a huge bond package and gave the city energetic leadership.[23] Indeed, one of their first acts was to claim through a first reading their intent to annex over seventy square miles of territory including much of the land between Grand Prairie and Dallas as well as areas north and south of the city. As a result, when Powell finally completed the city plan in October 1954, it anticipated a city of nearly one hundred square miles and a population of 500,000.[24]

In the introduction to the plan, Powell wrote that "Grand Prairie is not an isolated city, growing independently, influenced by factors peculiar to herself and in relation to the needs of a surrounding service area, as a City normally grows." But he went on to point out that this new type of city would still need strong leadership and a solid plan for success to take advantage of its position within the metropolitan area. Indeed, he concluded that "Grand Prairie can attract to herself practically any segment of that growth which she chooses to have, and at practically any rate she chooses to set." Although the plan continually emphasized Grand Prairie's close ties to the larger metropolitan community and called for "wise and farsighted cooperation with neighboring authorities toward an effective over-all development of the area," it also suggested the opportunity to make Grand Prairie a new kind of city, one that avoided some of the problems that nearby Dallas and Fort Worth experienced. After observing that "she *could* just grow as a residential suburb for the larger cities," the plan suggested the city could do more by providing ample land for industrial development and creating a business district to serve "the entire metropolitan area." The plan also called for "diversified development" so that "each prospective home-owner—or renter—can find the type of home and neighborhood which best suits his individual taste and means."[25]

Indeed, in anticipation of an expanding black population that was already at 641 according to the 1950 census, the plan called for the expansion of the black residential area now located in South Dalworth to the west.[26] It called for "a diversity of housing for the negro citizen," including "better types of homes, for those who desire and can afford them, as well as minimum cost housing." The plan also proposed "schools, parks, shopping districts and other neighborhood features" for the city's black residents. Not content with planning for the future, the plan also challenged Grand Prairie leaders to "improve or correct substandard conditions [existing] in the present development" and pass legislation to ensure that the established black neighborhood would "conform to minimum requirements for health, safety and convenience." It concluded by suggesting, "Grand Prairie should welcome and actively promote the development of a negro community which will be a show-place of the metropolitan area and a model for future development."[27] Such a recommendation to improve the

South Dalworth area might explain why Grand Prairie secured federal urban renewal money to improve that neighborhood in 1958.[28]

Grand Prairie's first plan followed a pretty traditional format in terms of employing the neighborhood unit plan for residential development and a grid street layout, but it did other things to make Grand Prairie a suburban city. For instance, it called for an 8,000-acre industrial district in the vicinity of Bear Creek and the development of downtown as both a "suburban shopping center" and a center for businesses "whose sphere of activity is the entire metropolitan area." More controversial was its call for thirty-foot building setbacks and the possible rerouting of the Texas and Pacific Railroad north of the city.[29] Such a plan clearly suggests that Powell had envisioned Grand Prairie not as a mere suburb but rather as a major player in the entire metropolitan region.

One of the areas that experienced the most rapid growth of housing tracts in the 1950s was the land in and around Irving, a community located just east of Dallas (driving time twelve minutes) and north of Grand Prairie (driving time ten minutes). That area benefited significantly from the nearby industrial growth in Dallas and Grand Prairie and the new building techniques that allowed rapid erection of plat housing on open land. In 1950 the two-square-mile community of Irving had a population of 2,621, but the eighteen square miles of unincorporated land surrounding Irving experienced rapid growth to a population of 7,500. The *Dallas Morning News* characterized Irving as "a vegetable-growing center and a sleepy commuter town before World War II," but that changed as developers Carr Collins and Pip Pipkins erected massive housing projects during the 1950s in and around Irving.[30] Mayor C. B. Hardee also contributed to the housing binge with his construction firm. Indeed, as early as 1948 *Times Herald* associate editor Bert Holmes called Irving "a suburb surrounded by its own suburbs." The *Architectural Forum* christened Irving an "instantaneous suburb," with "just homes, roads, cars, children and signposts promising more houses." Partially in response to the annexation efforts of Dallas and Grand Prairie, and partially because of the desire to control the growth just beyond the town's limits, Irving secured home rule status in 1953 and starting annexing and reserving subdivision tracts beyond the community's boundaries.[31]

Even before incorporation of Irving as a home rule city, the city commission created a planning commission, on March 11, 1952, and it recommended that the city employ a planning consultant, an idea strongly endorsed by the Greater Irving Chamber of Commerce.[32] After the city secured home rule, it employed the firm of Koch and Fowler to develop both a zoning ordinance and a city plan. City commissioners passed the zoning ordinance on December 22, 1953, and the consulting engineers completed the city's first city plan in 1954. Unlike the planning documents for Arlington and Grand Prairie, the initial plan for Irving seemed to be more for a suburban community of commuters since, according to the planners, Irving was primarily a residential community with more than 90 percent homeowners. Indeed, early on in their plan Koch and Fowler wrote, "It must be freely admitted at the outset that the community of Irving cannot boast self-sufficiency economic-wise", and must understand that "influences of a much larger area than Irving itself must be dealt with in relating local factors of economic stability and growth to those of the general area." Since more that 42 percent of the lands within Irving (taking up seven square miles) remained vacant, the planners recommended development within the community rather than annexation and development of further outlying areas. The sixty-page plan proposed a street and highway plan, provided for the development of accessible parks and playgrounds, and described land subdivision controls as well as other land-use regulations to create a "desirable and convenient environment" for its residents. In addition, it called for school and public utility improvements. The plan also underscored the importance of the Irving's central business district, arguing that it was "vital to the continuing economic stability of the city." In all, the planners provided "a comprehensive, coordinated document of related improvement and controls." Planning for industrial development was included in the plan, but on a much smaller scale and with less emphasis than in Arlington, Grand Prairie, or Garland.[33]

The limited commitment to industrial development by the plan did not mean that Irving city officials and civic leaders were not interested in industrial growth. Indeed W. H. Roberts, the president of the Greater Irving Chamber of Commerce, called for industrial expansion in the suburb after a presentation by Beryl Godfrey, former Fort Worth chamber of commerce president. Godfrey predicted that Irving would benefit from the rapid growth of manufacturing in the area, and civic leaders seemed to agree. The booster rhetoric from both the chamber of commerce and the young and energetic city officials suggested that Irving's leadership was not content with simply promoting their growing community as a bedroom suburb.[34]

Only Grand Prairie benefited from World War II more than Garland. Located east of Dallas, Garland became a center for airplane parts production as well as the manufacture of tanks. As a result, its population rose from 2,233 to 10,571, making this country town the third-fastest-growing community in Texas during the decade of the 1940s. Even before it became a home rule city, the mayor appointed its first planning board in April 1950 and asked it to draw up an overall plan to guide the city's growth.[35] After consulting with Carr Forrest and James Cotton, Dallas consulting engineers, the city approved zoning and devised a master street plan for the community's fourteen-square-mile municipality. Shortly after that, in response to Dallas's spatial expansion through annexation, Garland initiated steps to become a home rule city. Voters finally approved the proposal on October 16, 1951. Under the home rule charter, Garland appointed a new and more powerful city plan commission that turned to Forrest and Cotton for guidance.[36]

Unlike the other three cities examined here, Garland did not complete a comprehensive master plan in the early 1950s, but it did secure federal money in 1956 to undertake an urban renewal survey of the city's deteriorating Cooper-Barger area, the eastern part of Garland and home to many of the city's African American population.[37] Indeed, Garland became the second city in Texas (Grand Prairie was first) to receive a federal grant to study its slum and blighted areas. Although citizen opposition eventually thwarted plans to execute the urban renewal plan in the city's blighted areas, the council proceeded with urban renewal without federal help and appointed an eleven-member urban renewal committee to develop an alternative plan for financing slum clearance and housing rehabilitation.[38] Despite their focus on improving the city's blighted area, city officials understood the need for a comprehensive plan for the city during this period of rapid residential, commercial, and industrial growth. With the help of the federal government, city officials engaged the planning consultants Marvin R. Springer and Robert F. Foeller to produce a city plan in the early 1960s.[39]

The city did this with the help of the 701 Program, which provided matching grants "to state or metropolitan area government planning agencies to cover the cost of technical assistance for small cities and towns and for metropolitan regions within a state." This legislation sought to encourage a more ordered metropolitan region and combat what some saw as unplanned sprawl. In 1955, the first year of the program, Congress appropriated $1 million in grants for cities and counties under 25,000 population. By 1959, Congress had increased appropriations to $3,250,000 and made the grants available to communities with fewer than 50,000 inhabitants.[40] This program had a significant impact on greater Dallas not only because it encouraged smaller communities in the area such as Farmers Branch and Cedar Hill to engage planners but because it allowed the cities we have discussed to develop new and more detailed plans to cope with their massive growth.

In Garland, this program provided the necessary money to fund Springer and Foeller's master plan for the city, "Planning for Urban Growth."[41] As one of the original boom suburbs thanks to World War II, Garland continued to grow during the 1950s, going from 10,571 in 1950 to 38,501 in 1960, a 279 percent increase.[42] Although the plan offered guidelines on how to cope with the recent surge of population by exploring land use, thoroughfares and parks, schools, and neighborhoods, it also emphasized strategies for continued urban growth to ensure that Garland would remain *"the major suburban City* in the northeastern part of Dallas County." Just as the other plans had done, Springer's discussed expanding a well-balanced economic base as well as promoting appropriate land use. Even though the community already had substantial industrial development (more manufacturing jobs than residents), the planner proposed additional industrial areas along the city's two rail lines. Indeed, in his letter of transmission planner Marvin Springer noted that "one of Garland's most difficult growth problems will be the retention of adequate industrial area to permit a continued expansion of industry in the future."[43]

The plan also emphasized the importance of developing a strong central business district, since "without a central area, Garland would become, eventually, just a series of neighborhoods grouped around Dallas." This would not be easy, Springer acknowledged, since shopping centers took much of the commercial traffic away from the core. Still, the planner thought that specialty shops along with financial institutions, business offices, government offices, club headquarters, theaters, and restaurants would make a dynamic downtown. In addition, he proposed a system of thoroughfares that would provide easier access to the city's downtown.[44]

Although this plan encouraged cooperation with nearby communities and suggested that Garland's future was closely tied to the future of the entire metropolitan region, it also emphasized ways that Garland could achieve a competitive edge over other nearby communities and stressed the importance of Garland becoming a self-sufficient city. The title of the plan, after all, was not "Planning for Suburban Growth" but rather "Planning for Urban Growth."

Grand Prairie continued to grow during the 1950s and 1960s. Helped by the development of a major industrial park called the Great Southwest Industrial District, a 5,000-acre area shared with Arlington, Grand Prairie boosters predicted that their city would provide "the most fantastic scene of industrial development this country has ever known."[45] To prepare for that growth, the city also embraced additional planning thanks to the 701 Program. It received federal planning funds in 1957 to develop a street plan. Two years later it secured additional money from the federal government's 701 Program to employ Homer Hunt and Associates, Dallas consulting engineers. Hunt made a complete land-use study of the city and the lands under first annexation reading, revised the zoning ordinance, and updated the master plan of 1954.[46] Several years later, Grand Prairie employed Springer and Associates as planning consultants, and that resulted in a comprehensive plan report for the city of Grand Prairie in 1966. That plan, like earlier ones, emphasized the need for Grand Prairie as a suburban city to stay "competitive with all other communities in the area as a place in which to live, earn a livelihood and rear a family."[47]

Irving, the community that had been characterized as a typical bedroom suburb as late as 1961, secured federal funds in May 1958 to promote additional planning. Such money would allow the community to plan and control its own future development by "making constructive decisions" about its future development, according to planner Hugo Leipziger. City officials secured the services of Hugo Leipziger-Pearce and Associates in August 1958 to produce a new comprehensive plan for the community, which was the fastest-growing one in Texas during the 1950s, expanding from 2,621 in 1950 to 45,489 in 1960, a 1,736 percent growth rate.[48] Leipziger, head of the Department of Planning and Architecture at the University of Texas, actually drew up seven preliminary reports in addition to the final comprehensive plan completed in July 1960. His plan included not only the city itself but also the land held for annexation beyond city limits by a first reading.[49]

Like the other suburban cities, Irving grew not only because of the outward movement of population from Dallas but because it undertook aggressive annexation of its own outlying areas, including those large housing tracts built just beyond its city limits.[50] Although much attention was given to promoting "residential excellence" through embracing the neighborhood unit concept, this plan called for more. After acknowledging the city's limited industrial base (82 percent of Irving's labor force worked outside the city), the plan proceeded to suggest that "Irving should broaden its present dormitory character into a better-balanced pattern of residential, commercial, and industrial development." As a result, it called for more land dedicated to industrial use but within the goal "of keeping Irving an attractive residential community." Such action, according to Leipziger, would "strengthen the competitive advantage of Irving over other communities in the Dallas-Fort Worth area."[51]

As with Garland and Grand Prairie, the Irving plan also provided attention to parks and playgrounds, schools, and thoroughfares and emphasized the importance of maintaining a strong and vital downtown. Indeed, the Irving plan called for a "tightly knit entity with a full range or retail, administrative, financial, service and entertainment facilities." Furthermore, retail needed to be preserved against shopping center competition by adding pedestrian malls and adequate parking. Above all, the plan concluded, Irving's downtown should not be allowed to deteriorate.[52]

Arlington also revised its plan, in both 1959 and 1964, with both revisions by Robert Caldwell. Like the other city plans, these not only emphasized residential development but envisioned the city as "an outstanding industrial, commercial and cultural center." It predicted that Arlington would continue to grow "as an individual urban entity and as the axis and vital center of the ever growing Dallas-Fort Worth Metropolitan Area." Although the planner classified Arlington at that time as "principally a residential City," with three-fourths of the working population commuting outside its corporate limits, he predicted that would change with rapid industrial and commercial growth as North Texas continued to expand. Indeed, by the time of the 1959 plan, Arlington not only had the GM plant but also included the massive Great Southwest Industrial District shared with Grand Prairie.[53]

As with the other cities, there was some interest in maintaining the central business district, but the plan acknowledged that Arlington's district could not compete with Dallas and Fort Worth "as a major market or commerce center." It also concluded that the appearance of outlying shopping centers was sapping the vitality of downtown. Still, both the 1959 and the 1964 plans suggested that the city's core be more a center of government, financial, and professional uses rather than retail.[54]

Although Arlington had a different look than the traditional city, all three plans for that community envisioned creating a discrete unit that was much more than a bedroom community. Indeed, the 1964 plan made this point by reminding its readers, "It is necessary to study the Arlington urban area in its capacity as a growing major urban area, rather than a village of yesteryear." Although the plan continued to emphasize that Arlington was a "city of homes," it also called for more multifamily unit construction and encouraged the city to continue to grow "its own industrial base."[55] The 1964 Arlington plan was a blueprint for a city located within a huge metropolitan area rather than a suburb completely dependent on the nearby cities of Dallas and Fort Worth.

A close look at the planning efforts of these four North Texas suburbs suggests that they were not simply "accidental cities" but intentional communities nurtured by their energetic civic leadership, shaped by their location within a growing metropolitan area, and formatted by professional urban planners hired by local booster governments to create what became a new urban form—suburban cities. Despite an initial popular perception of these places as nothing more than bedroom suburbs,[56] it became evident shortly after the war that some civic leaders of those places had other ideas. They were not content with small, intimate, homogeneous communities but saw some economic and social advantages to promoting a new type of place.[57] All four cities annexed nearby land aggressively, trumpeted the locational advantages of their places to anyone who would listen, and engaged city planning experts to create a new type of urban place. Some, like Grand Prairie and Garland, had a head start because of World War II; others, like Arlington and Irving, followed the recommendations of their planners and sought out industry and a much greater land mix than found in a dormitory suburb to pursue their own type of urban dreams. Not all plans were followed to the core; for instance, Grand Prairie refused to push for rerouting of the main line of the Texas and Pacific rail tracks bisecting their city.[58] Nevertheless, the general planning principles of segregating land use, following the neighborhood unit plan, providing parks, preserving the old downtown, and developing road systems to make the city more acceptable clearly shaped these suburban cities. These cities also welcomed blue-collar as well as white-collar workers to live and work within their city limits and even seemed willing to allow black migration as long as blacks stayed in their segregated neighborhoods.[59]

Even though the 701 Program focused on promoting a more ordered and coordinated metropolitan area, by the 1960s suburban cities had used federal funds to better their own position in the metropolis even if it came at the expense of other communities within the larger region. Some even turned to urban renewal money to improve the lot of their neediest residents. Although these places were willing to cooperate with each other at times, the greatest emphasis of their plans appeared to focus on promoting self-sufficiency and becoming independent urban entities. Just as in cities a century earlier, urban boosterism and suburban city rivalry ruled the day as civic leaders vied to make their communities significant players within the metropolitan region.

NOTES

1. See, for instance, the collection of essays in Kevin M. Kruse and Thomas J. Sugrue, eds., *The New Suburban History* (Chicago: University of Chicago Press, 2006).

2. North Central Texas Council of Governments, "Population Estimates," www .nctcog.dst.tx.us/ris/demographics/population.asp (accessed May 18, 2010).

3. Robert Fishman, *Bourgeois Utopians: The Rise and Fall of Suburbia* (New York: Basic Books: 1987); Joel Garreau, *Edge Cities: Life on the New Frontier* (New York: Doubleday, 1991); Carl Abbott, *Metropolitan Frontier: Cities in the Modern American West* (Tucson: University of Arizona Press, 1993); Jon C. Teaford, *Post-suburbia: Government and Politics in the Edge Cities* (Baltimore: Johns Hopkins University Press, 1987).

4. Robert E. Lang and Jennifer B. LeFurgy, *Boomburbs: The Rise of America's Accidental Cities* (Washington, DC: Brookings Institution Press, 2007).

5. See, for example, William H. Whyte Jr., *The Organization Man* (New York: Simon and Schuster, 1956); Ann Forsyth, *Reforming Suburbia: The Planned Communities of Irvine, Columbia and the Woodlands* (Berkeley: University of California Press, 2005); and Zane L. Miller, *Suburb: Neighborhood and Community in Forest Park, Ohio, 1935–1976* (Knoxville: University of Tennessee Press, 1981).

6. Walter S. Robinson, "Suburbia," *Dallas Morning News*, January, 20, 1963, sec. 10, 1; "Master Plan Study Slated for Mesquite," *Dallas Morning News*, November 4, 1959, 9; Marvin R. Springer and Associates, *A Report on the Comprehensive Plan for Richardson, Texas* (Richardson, TX: Marvin R. Springer and Associates, 1962); Jim Key, "Plano Ready Now with Plans for Expected Huge Growth," *Dallas Morning News*, January 20, 1957, 7. Robinson mentions fourteen suburbs, and with Arlington, Grand Prairie, Irving, Garland, Plano, and Richardson that runs the number to twenty. Robinson's fourteen suburbs are Mesquite, Coppell, Carrollton, Farmers Branch, Addison, Sunnyvale, Balch Springs, Kelberg, Seagoville, Hutchins, Lancaster, DeSoto, Cedar Hill, and Duncanville.

7. "Turner Says 24 Years Enough," *Dallas Morning News*, February 20, 1949, 1.

8. Walter Robinson, "Plane Plant Town Booms," *Dallas Morning News*, July 16, 1950, sec. 4, 1.

9. "Texas Almanac: City Population History from 1850–2000," *Texas Almanac*, http://texasalmanac.com/sites/default/files/images/CityPopHist%20web.pdf (accessed May 27, 2016).

10. For instance, by 1948 Grand Prairie had sixty industrial plants and Arlington was now home for twenty-three industries. "Big Boom for Grand Prairie Still to Come, Boosters Say," *Dallas Morning News*, May 22, 1948, 4; "Arlington Boosts 23 Industries but Puts Emphasis on Homes," *Dallas Morning News*, May 22, 1948, 2.

11. "Arlington's Master Plan to Be Prepared by Freese and Nichols," *Arlington Journal*, October 5, 1951, 1; Freese and Nichols, Consulting Engineers, Robert W. Caldwell, Associate Planner, *The City Plan for Arlington, Texas, 1952: A Guide for Future Development* (Forth Worth, TX, 1952), 9–11.

12. Allan Saxe, *Politics in Arlington Texas: An Era of Continuity and Growth* (Austin: Eakin Press, 2001), 24–25. See also David Lynn Cannon, "Arlington's Path to Post Suburbia," Ph.D. dissertation, University of Texas at Arlington, 2000.

13. Guest lecture by Tom Vandergriff, History 3351, University of Texas at Arlington, April 26, 2000.

14. "Vandergriff Elected Mayor; Gee, Wolf Voted to Council," *Arlington Journal*, April 6, 1951, 1; "Mayor Vandergriff, New Commission Sworn into Office," *Arlington Journal*, April 13, 1951, 1.

15. "New City Planning Can Be a Big Thing for Arlington," *Arlington Journal*, June 29, 2; "A Master City Plan Can Help Community Avoid Headaches," *Arlington Journal*, September 21, 1951, 10; "Zoning Commission Set Up by City," *Arlington Journal*, January 19, 1952. Quote from Freese and Nichols, *City Plan for Arlington*, 20.

16. "We Must Zone for All, Not One," *Arlington Journal*, March 14, 1952, 2; "City Zoning Ordinance," *Arlington Journal*, April 25, 1952.

17. Terri Myers, *The Hill: Arlington's African-American Communities, a Report on the Growth and Development of Arlington's Historic African-American Communities, 1845–1999* (Austin: Hardy, Heck, Moore and Myers, 1999), 2–4.

18. Fresse and Nichols, *City Plan for Arlington*, 95. The five-block area had an assortment of shotgun houses and small bungalows, some with lots large enough for gardens. According to its historian, the Hill was "socially and physically separated from [white] Arlington with neither paved roads, curbs or gutters until the mid-1960s. Home ownership was moderately high as men worked as day labor or held janitorial jobs at local industry while women worked as hired help in white homes" (14–20).

19. "City Zoning Ordinance," *Arlington Journal*, April 25, 1952.

20. "City Plan to Be Presented Tuesday," *Arlington Journal*, September 19, 1952, 1; Arlington Plan Given to Community," *Arlington Journal*, September 26, 1952, 1; Freese and Nichols, *City Plan for Arlington*, 9–11.

21. Powell and Powell, *City Plan for Greater Grand Prairie*, 1.1, 3; "Dallas Annexes Industrial Area, Pushes Limits to Grand Prairie," *Dallas Morning News*, September 7, 1947, 1; "Grand Prairie Voters Decide on Home Rule," *Dallas Morning News*, May 2, 1948, 1; William Neil Black, "Empire of Consensus: City Planning, Zoning, and Annexation in Dallas, 1900–1960," Ph.D. dissertation, Columbia University, 1982, 285–89.

22. Kathy A. Goolsby, *A History of Grand Prairie pre 1840s–2009* (Grand Prairie, TX: City of Grand Prairie, 2013), 75–78; Letter of transmission from Powell and Powell, Consulting Engineers, Dallas, in Powell and Powell, *City Plan for Greater Grand Prairie*, chap. 1, 1.19.

23. Walter S. Robinson, "Whiz Kids of Grand Prairie City Council Making History," *Dallas Morning News*, October 11, 1953, 1. Youthful enthusiasm worked against the "whiz kids," especially their aggressive annexation plans, which resulted in numerous lawsuits against the city. Businesspeople and boosters developed a new ticket called "Grand Prairians for Good Government," to run against the group in 1955, and it swept into office. That group was responsible for the establishment of a city manager and a more professional city administration. Goolsby, *History of Grand Prairie*, 86.

24. *Dallas Times Herald*, May 7, 1953; Powell and Powell, *City Plan for Greater Grand Prairie*, 1.2.

25. Powell and Powell, *City Plan for Greater Grand Prairie*, 1.1–2, 1.40, 1.69.

26. U.S. Census Bureau, *Population Census, Texas Places, for 1950*, 43–101. The city annexed Dalworth Park in 1943, a community that included a significant black population. Lisa C. Maxwell, "Grand Prairie, TX," *Handbook of Texas Online*, www.tshaonline.org/handbook/online/articles/HDG03 (accessed May 26, 2016).

27. Powell and Powell, *City Plan for Greater Grand Prairie*, 1.5–6.

28. "Renewal to Begin in Grand Prairie," *Dallas Morning News,* June 8, 1958, 5.

29. Powell and Powell, *City Plan for Greater Grand Prairie,* 1.7, 1.32–33.

30. Walter Robinson, "Irving Riding a Crest of Boom," *Dallas Morning News,* February 17, 1952, 1; Joseph Rice, *Irving: A Texas Odyssey* (Chatsworth, CA: Windsor, 1989), 71–72; Hugo Leipziger-Pearce and Associates, *Preliminary Development Plan for Irving, Texas, 1960* (Austin: Hugo Leipziger-Pearce and Associates, 1959), 6.

31. Holmes quoted in *Dallas Times Herald,* April 28, 1974; "Freeway Suburb," *Architectural Forum* 114, no. 1 (January 1961): 76; Terrell Blodgett, *Texas Home Rule Charters* (Austin: Texas Municipal League, 1994), 161; "Irving to Annex 18 Square Miles," *Dallas Morning News,* February 21, 1953, 6.

32. "Master Plan Development Backed by Irving C of C," *Dallas Morning News,* March 25, 1953, 18.

33. Koch and Fowler, Consulting Engineers, *The Irving City Plan, Irving Texas, 1954: A Guide for Future Development* (Irving, TX: Koch and Fowler, Consulting Engineers, 1954), 1–6, 11–12, manuscript vertical file, Irving Archives in the Irving Public Library, Irving, Texas. Oscar H. Koch and James D. Fowler headed a well-established engineering firm that had done a plan for Austin in 1928. Koch had earlier been the public works engineer for Dallas, and the consultants had done some road planning for Irving in 1950. Koch and Fowler, Consulting Engineers, *A City Plan for Austin, 1928* (second printing, Austin: Department of Planning, City of Austin, 1957); "Firm Hired to Manage Road Work, *Dallas Morning News,* February 12, 1950, 1; "Irving Draws Up Zoning Ordinance," *Dallas Morning News,* October 29, 1953, 8.

34. Walter S. Robinson, "Future Industry Gains Told to Irving C of C," *Dallas Morning News,* February 20, 1953, 1; "Boom City Irving a Pasture in 1903," *Dallas Morning News,* October 18, 1953, 1. Even the *Dallas Morning News* had to acknowledge the new commitment of Irving to become more than a commuter suburb. In an editorial it admitted, "Irving folk, as usual, are on their toes. Not satisfied to grow bigger from the overflow of Dallas, they are taking steps toward making theirs a better integrated community." "Irving Master Plan," *Dallas Morning News,* March 31, 1953, 2.

35. "Garland Picks Plan Leader," *Dallas Morning News,* April 26, 1950, 9.

36. "Plan Board Announced by Garland," *Dallas Morning News,* December 20, 1951, 12; Michael R. Hayslip, *Garland: Its Premiere Century* (Chatsworth, CA: Windsor, 1991), 66.

37. "3-Way Pact Studied for Renewal Work," *Dallas Morning News,* August 31, 1958, 11; "Slum Work Given OK," *Dallas Morning News,* October 25, 1956, 3. According to the census bureau, Garland's black population increased from 436 in 1950 to 1,465 in 1960. U.S. Census Bureau, *Population Census, Texas Places,* for 1950, 43–101; U.S. Census Bureau, *Population Census, Texas Places* for 1960, 45–119.

38. "Garland Drops Plan to Ask Renewal Aid," *Dallas Morning News,* June 4, 1959, 6; "UR Program Planned without Federal Help," *Dallas Morning News,* October 25, 1959, 17.

39. The two men formed the firm of Springer and Foeller, Urban Planning and Area Development Consultants, in July of 1960. "In Dallas," *Dallas Morning News,* July 21, 1969, 18.

40. Carl Feiss, "The Foundation of Federal Planning Assistance: A Personal Account of the 701 Program," *Journal of the American Planning Association* 51 (Spring 1985): 181–82.

41. "Garland May Employ Springer as Planner," *Dallas Morning News*, August 13, 1959, 16.

42. Marvin Springer and Associates, *Planning for Urban Growth, Garland Texas* (Dallas: Marvin Springer and Associates, 1962), 7.

43. Marvin R. Springer to Honorable Mayor, City Council, letter of transmission, February 19, 1962, in Springer, *Planning for Urban Growth*.

44. Springer, *Planning for Urban Growth*, 86–87.

45. "Grand Prairie Now Ready for 'Fantastic' Developing," *Dallas Morning News*, January 19, 1958, 14.

46. Allen Quinn, "Small City Feels Pains of Growth," *Dallas Morning News*, June 27, 1957, 1; "Town to Get Federal Plan on Master Plan," *Dallas Morning News*, September 3, 1959, 10; "3-Way Pact for City Survey," *Dallas Morning News*, October 1, 1959, 21.

47. Springer and Associates, *Comprehensive Plan Report: City of Grand Prairie* (Dallas: Marvin Springer and Associates, January 1966), A.

48. "Federal Grant to Pay Half of Municipal Plan," *Dallas Morning News*, May 23, 1958, 3; *Irving News*, November 6, 1958; Leipziger-Pearce, *Preliminary Development Plan*, 3.

49. "Irving to Draw Up City Plan," *Dallas Morning News*, June 22, 1958, 12; *Irving News*, October 10, 1958; Leipziger-Pearce, *Preliminary Development Plan*, 22.

50. Leipziger-Pearce, *Preliminary Development Plan*, 6. Irving went from two to twenty square miles during this period and reserved another fifty square miles by a first reading of annexation.

51. Leipziger-Pearce, *Preliminary Development Plan*, 23, 40, 72.

52. Leipziger-Pearce, *Preliminary Development Plan*, 77.

53. Caldwell and Caldwell, *Arlington City Plan: Studies and Revision, 1959* (Arlington: City Zoning and Planning Commission), 1, 7–8.

54. Caldwell and Caldwell, *Arlington City Plan, 1959*, 26; Robert W. Caldwell and Associates, *Comprehensive Master Plan, 1964: Studies and Revisions. A Guide for the Development of Arlington, Texas* (Arlington: Robert W. Caldwell and Associates, 1965), 132.

55. Caldwell, *Comprehensive Master Plan*, 3, 133–34.

56. The best characterization of this in North Texas is the article "Freeway Suburb," *Architectural Forum* 114, no. 1 (January 1961): 76. It pointed out that "Irving tragically represents too many other galloping suburbs across the U.S., suburbs which have inflated the worst possibilities of their native landscapes."

57. In his excellent study of the origins of post-suburbia, Jon Teaford argued that the adoption of commerce and industry was a compromise to keep suburbia financial vital and not a strategy to create a new type of city. My analysis suggests that it was more than just that. Jon C. Teaford, *Post-suburbia: Government and Politics in the Edge Cities* (Baltimore: Johns Hopkins University Press, 1997), e.g., 8.

58. Walter S. Robinson, "Master Plan Stirs Row in Suburb," *Dallas Morning News*, June 13, 1954, 1.

59. This North Texas experience seems to challenge or at least call into question the conclusion of William Sharpe and Leonard Wallock that suburban cities were really cities based on cultural criteria rather than a functional approach. William Sharpe and Leonard Wallock, "Bold New City or Build-Up 'Burb?" *American Quarterly* 46 (March 1994): 17–23.

{ 3 }

SUBURBAN STRIFE AND THE PERILS OF MUNICIPAL INCORPORATION IN TEXAS

ANDREW C. BAKER

Shaking the dust off of their feet after World War II, Americans left the city for life in the countryside—or at least in housing developments that promised a countryside lifestyle. These self-identified frontier families ventured out past the city limits, buying into a new world developers had created to appeal to their consumer tastes. They left the city and its problems far behind. As they settled into their new homes and neighborhoods, however, many found that they had taken for granted the municipal services that supported their modern, American standard of living. Moving across the city line meant moving into counties that were often incapable of providing clean water, sewer, paved roads, drainage, trash removal, building codes, and police and fire protection. These families quickly realized that their suburban dream would require the work of lawyers, state agencies, local governments, and residents as much as it had already depended on the work of surveyors, county clerks, developers, and home builders. Constructing suburbia required the extension of state structures into the countryside and the strengthening of county governments to approximate more fully the municipal governments residents had left behind. When these newcomers ripped off the wrapping paper from their suburban paradise, they found it came packaged in a box labeled "some assembly required."

Daily confronted with their unmet expectations, these homeowners, and the developers who sold to them, chose among four available solutions. One option was to expand the reach and powers of county governments to fill the gap in services. A second option was to create special taxing districts approved by the state, to provide specific services within limited boundaries and paid for exclusively through taxes levied on the service population. A third was to submit to—or even request—annexation by the city, tying these properties into urban infrastructure at the cost of local political control and community identity. The final option was to incorporate—to organize and create a suburban town with many of the powers and responsibilities of the cities they had left behind.

Which option residents chose shaped their settlement's subsequent history at least as much as did the developer's particular sales strategy, the layout and quality of the housing, or the quality of the natural surroundings. The local political strife surrounding these decisions could be just as contentious as debates over taxation, security, or desegregation. In fact, it often contained aspects of all three. Yet this question of how to provide community services has often been overshadowed in the historiography by an emphasis on suburban image and national politics. Whereas Kenneth Jackson's *Crabgrass Frontier* focuses extensively on incorporation, annexation, and the difficulties of providing services to suburban areas, later treatments have marginalized these issues in their focus on racial politics, cultural values, architecture, and environmentalism. In the process, the historiography has lost sight of much of the rich history of grassroots conflict involved in making suburbs.[1] This chapter sketches these four general options as they played out in Texas. Here, as in most states, suburbs negotiated between city, county, and special district—between annexation, incorporation, and privatization. In Texas, a Sunbelt state, the rules of the game were written in the city's favor. Aggressive annexation became the norm between the late 1940s and the early 1980s—a fact with profound implications for the state's suburbs.

This chapter focuses on the history of one of these subdivisions. For residents of Chateau Woods, a small rural subdivision in Montgomery County north of Houston, the threat of annexation by Houston propelled a movement to incorporate this struggling community. They succeeded in 1976, removing that threat once and for all. They protected themselves from *the* city by creating a new city under their control. Many recognized their mistake soon after. The town of Chateau Woods struggled to cope with the responsibilities of incorporation and ultimately failed as a general law city. In addition to demonstrating the frustrations of incorporation and the limits of suburban politics, this case study opens a window into the local impact of Reagan-era political rhetoric. The history of Chateau Woods reveals fundamental disagreements among residents over whether suburban developments should be civically engaged communities run by elected officials or fiscally responsible businesses run by interested parties; whether they comprised citizens, neighbors, or homeowners; and whether they should be led by housewives or businessmen. These unresolved conflicts lie at the heart of suburban politics.

Texas county governments were the most obvious vehicle for meeting suburban needs, yet time and time again they proved unable to provide either the type or level of services newcomers expected. Legally, counties derived their powers from the state and therefore could exercise only powers expressly granted by the Texas legislature. County commissioners' courts handled road construction, repair, and drainage, oversaw tax assessment, and coordinated with the offices of the county sheriff, county attorney, and agricultural extension. Landfill operations, building standards, health and safety codes, and flood regulations each required enabling legislation from the state. Even as the legislature gradually opened these avenues in response to mounting pressure from metropolitan counties, many local governments were hesitant to take on the additional costs, logistical headaches, and political retribution that often came with offering improved services and increased regulation paid for by increased taxation. Those services counties did offer were generally underfunded, overextended, and unprepared for the challenges of metropolitan development. Newcomers looking to the county for services, therefore, faced an uphill battle against both entrenched rural political power structures and state limits on the legal powers available to county government.[2]

Most developers preferred to establish service districts to provide services to their customers. Nationally, the number of these districts grew rapidly, from fewer than five hundred in 1962 to ten thousand by 1970 to twenty-thousand by 1975.[3] The municipal utility district (MUD) became the most popular form in Texas after the legislature created it in 1971, eclipsing fresh water supply districts, conservation and reclamation districts, and flood control districts. MUDs could manage trash collection and everything related to the movement of water, from irrigation and drainage to treatment and supply. During the first year, Texans created 110 MUDs. By 1979 there were 392 districts in the Houston metropolitan area alone and 1,028 in the state.[4] Montgomery County approved close to a dozen MUDs during the first year and forty-two by spring 1978.[5]

Creating a MUD was rather simple. Those owning a majority of the value of land within a proposed district petitioned the state for its creation.

Once approved, the district went up for election. Approval required a majority of voters within its boundaries. In practice this often meant that a handful of property owners friendly to the developer approved the district before the developer opened lots to homebuyers. The districts were generally approved, established, and funded through bonds before residents moved in. At that point, the developer transferred control of the district, with its assets and liabilities, to the residents.[6]

Service districts have had their share of critics. At the metropolitan level they tend to reinforce and perpetuate the fragmentation of services and civic engagement, promote a consumer mindset toward utilities, and undermine economies of scale. At the local level, voter apathy and ignorance allow a small group of professionals to manage these entities with little oversight.[7] Yet, whatever their limits, such districts served developer interests and therefore became the most common solution to the pressing service needs of subdivisions outside city limits.

City limits, though, did not stay put. Urban territorial expansion through municipal annexation posed a constant threat to suburban self-determination. A number of factors encouraged cities to press their boundaries ever-outward. Annexation captured additional tax revenue. It boosted the statistical markers of a city's national and regional significance. It also allowed cities to regulate and direct surrounding development. These benefits generally outweighed the costs of extending city services, provided the city annexed only the choicest areas. Cities in the Northeast and Midwest, with the blessing of state legislatures, annexed aggressively from the mid-nineteenth century into the first decades of the twentieth. By the 1920s, however, many were being hemmed in by the proliferation of elite garden suburbs. These suburban enclaves turned to defensive incorporation to protect their social homogeneity and to ensure local control over development and tax revenues. Such incorporations would be the norm up through the 1980s.[8]

Nationwide, postwar annexation occurred at a rapid pace. Between 1970 and 1977, for instance, the nation averaged 6,013 annexations per year, accounting for an average total of 862 square miles. Most of these were small, averaging only 91.8 acres each.[9] This postwar annexation was largely a Sunbelt phenomenon. Here permissive state laws allowed the region's booming postwar metropolises to keep pace with their sprawling highways and subdivisions. During the 1970s, for instance, 40 percent of all annexation in the nation occurred in California, Florida, Texas, and

Illinois.[10] Texas was home to some of the most liberal annexation laws in the nation. The 1912 home rule amendment to the Texas constitution enabled cities with more than 5,000 residents to form a home rule city and unilaterally annex an unlimited amount of adjacent, unincorporated territory. The legal mechanisms for this annexation were left up to each city's charter.[11] The state supreme court expanded these powers even further, ruling that the principle of "first in time, first in right" applied to municipal annexation. This allowed home rule cities to complete their first legal reading of annexation proceedings for a given piece of territory without consummating that annexation. This first reading prevented any other city from annexing that territory and prevented communities within that territory from incorporating. In effect, cities could reserve territory for themselves, without the obligation to provide any services to them and without gaining tax revenues from them. Texas cities exercised powers over surrounding communities that were some of the strongest in the nation.[12]

Texas cities used these powers to their fullest, stretching the lines of their authority outward and driving their population statistics upward. San Antonio's city manager pursued a bold annexation program to capture tax revenues from what he termed "parasite cities" outside the city's northern limits. These actions pushed the city from 69 square miles in 1950 to 160.5 by 1960 but failed to capture the incorporated suburbs of Olmos Park, Alamo Heights, and Tarrell Hills.[13] In Houston, city leaders became embroiled in their own battles over annexation beginning in the late 1940s.[14] Word of secret meetings between the incorporated communities of Galena Park and Pasadena stoked fears among Houston city leaders that the city might be hemmed in if these municipalities chose to annex surrounding territory. Acting to avert this nightmare scenario, the city council acted in 1949 to expand the city's boundaries beyond these communities, doubling the size of the city and encircling the incorporated communities of West University, South Side Place, Bellaire, Galena Park, Jacinto City, and part of Pasadena.[15] In 1956, Houston further annexed nearly 185 square miles, much of it newly developed land. In the process it took control of thirty water control and improvement districts, sixty utility companies, and a bonded debt of $31 million—debt the city was now responsible for. This land included the future airport site and the city's Lake Houston reservoir. City leaders saw annexation of these territories as vital to regulating water, sewer, and drainage in these areas. They also believed the city's continuing spatial expansion was necessary to securing its economic future.[16]

As these examples suggest, annexation was a statewide trend. The state was home to five of the twelve cities in the nation annexing more than fifty square miles during the 1950s.[17] Smaller cities joined in. Some 105 municipalities in the state completed 2,010 separate annexations during the decade, adding 1,138.1 square miles to urban Texas. Twenty-one of the twenty-eight largest annexations were in metropolitan areas. The state was second in the nation, behind California, in the number of annexations involving more than fourteen square miles (966 annexations) and first in land area annexed and in population annexed (573,077 people). Some 20 percent of the people in the nation annexed during the 1950s lived in Texas.[18]

This "annexation fever" came to a head in 1960 in southeast Harris County. The smaller municipalities of Pasadena, Lomax, Deer Park, and LaPorte attempted to beat Houston in the race to secure tax revenues from the booming Clear Lake area. Their city leaders met secretly in Dallas to organize a coordinated preemptive strike on the city. On June 6, these cities claimed some 105 square miles east and southeast of Houston. The four municipalities were thereby poised to control one-quarter of Harris County. Houston leaders were blindsided. With expansion by Texas City and Baytown also looming, the Houston city council conducted its first reading of an ordinance to annex all unincorporated land in Harris County—a move that one journalist labeled "a burlesque of legitimate city growth" and Mayor Lewis Cutrer decried as ludicrous. This action set the city on a path to reaching 1,560 square miles, potentially making it the world's largest city by area. These rapidly cascading annexation moves, along with the howls raised by business leaders over the tax increases that they would bring, convinced the state legislature to step in and appoint a committee to chart a way out of this annexation mess.[19]

By August the fever had spread to North Texas. With legislative restrictions likely on the way, municipalities raced to lock in room for future expansion. Dallas, a city whose annexations over the past twenty years had expanded its area from twenty-five square miles to 283, found itself hemmed in by the county's twenty-seven municipalities. Irving, Denton, Plano, and McKinney were the most active. The legislative committee continued to hold hearings into the fall before presenting its findings. Although it found that close to half of the state's annexations were invited by residents, the scale and hostility generated by hostile annexations demanded action.[20] Houston city leaders warned of the threat that expanding incorporated communities posed to the economic health and orderly growth of the state's metropolises. Strong annexation powers, they believed, in spite of recent events, were needed to avoid metropolitan fragmentation and stagnation.[21] Some, like former Dallas planning director Marvin Springer, warned that any action to take away urban annexation powers would leave the state with "the hodge-podge of the Eastern cities— not the sound and orderly growth of Texas." Yet it was precisely the failure to keep growth "sound and orderly" that forced the legislature to pass the Municipal Annexation Act of 1963.[22]

The new rules required cities to complete annexation proceedings within ninety days after initiating them and limited them to annexing only 2 percent of their land area per year with the provision that they could bank each year's quota, up to 30 percent. This protected larger cities from the threat of small municipalities annexing large swaths of metropolitan land and provided some predictability to the process. A city annexing territory was now required to provide urban services to new territory within three years. If it failed to do so, residents could vote to de-annex themselves. The law's most important innovation was the extra-territorial jurisdiction (ETJ), a type of municipal frontier extending five miles beyond the boundaries of cities with over 100,000 residents, incrementally declining in size for cities with smaller populations. Within this territory the city could prevent incorporation and annexation as well as regulate and enforce codes. What they could not do was tax within their ETJ.[23]

Even with these legislative restraints, Texas cities continued to expand. The new law transformed the massive land grabs of the 1950s into predictable races between cities to extend their ETJ.[24] Strip annexation was the most popular method, wherein a city would annex a fifty-foot-wide ribbon of land through the countryside, extending its ETJ five miles on either side. Houston, for instance, began expanding its ETJ northward toward Montgomery County through strip annexation in 1965.[25] Civic leaders within the county seat of Conroe, angered by what they saw as encroachment by their "big brother" to the south, began annexing strips of land southward along I-45, defending central Montgomery County and the future Lake Conroe from the city's reach. This clash left the southern boundary of the county, where Chateau Woods was located, caught between encroaching ETJs.[26]

Even with these new limitations, the legislature bequeathed annexation powers to Texas cities that were far beyond those available to cities in the Northeast or Midwest. With the creation of ETJs, the metropolitan fringe became a battleground where cities and towns each maneuvered their boundaries and therefore their ETJs to consolidate their territory, regulate future growth, and prevent outside encroachment, often with as little actual annexation and therefore expense as possible. Many residents saw their suburban independence gobbled up into city limits, subsumed into a neighboring town, or locked into an ETJ-defined subservience. They feared that the city that they had fled would find them, claim their tax revenues, neglect their services, and marginalize them within the cacophony of urban politics.

MUDs and other such special districts could meet the immediate service needs of suburban communities. They had no power, however, to prevent annexation or the expansion of ETJs. The only complete defense was incorporation. This was an option whose implications most voters did not fully understand. Forming a municipal government gave residents the legal tools to protect the homogeneity and the character of their community, to prevent substandard development, to ensure their taxes paid for their services, and to apply for federal grant money. Yet these benefits came with significant costs.[27] Forming a competent government was no easy task. Incorporated subdivisions passed and enforced building codes and zoning; provided police and fire protection, and trash removal; maintained roads and managed stray animals; and provided water, sewer, and drainage for their residents. Residents, in turn, paid for these services through municipal taxes on top of their county taxes.

Incorporation provided a stronger community identity and the forms of local democracy, yet, even so, many of these communities continued to struggle against the weight of local apathy and limited resources. Those who thought they had left behind the city's rules, regulations, and constraints by moving to the metropolitan fringe found that a local town hall could be just as coercive and obstructive as city hall. The city hall brought with it the added frustration that the conflicts often turned personal and almost always contained a large dose of amateurism. Rejecting *the* city required creating *a* city.

"Annexation fever" set off a wave of incorporations in Texas. By 1960, Harris County had twenty-four municipalities. Dallas County had twenty-eight. The Houston, Fort Worth, Dallas, and Beaumont–Port Arthur areas combined had some ninety-seven municipalities. San Antonio's 1952 annexation push alone drove some sixteen communities to begin incorporation proceedings.[28] The numbers continued to mount. Between 1970 and 1977, Texas added 137 new municipalities. Houston and Conroe's race to extend their ETJs into southern Montgomery County in 1965 set off a chain reaction of incorporations in that particular region.[29] Caught next to an expanding city and faced with the perennial failures and inefficiencies of county governments and service districts, close to a dozen of the county's subdivisions incorporated over the following decade.[30] Some, like Panorama Village, Oak Ridge North, and Shenandoah, which secured committed and skilled leaders and a commercial tax base, survived and often thrived.[31] Others were not so fortunate. Whispering Oaks, for instance, struggled for years to cope with frequent flooding, ineffective septic tanks, and insufficient water treatment. The Federal Emergency Management Administration (FEMA) eventually bought most of the homeowners out and donated the land to the county, which allowed it to return to forest.[32]

Though other incorporated subdivisions did not fail quite so spectacularly, the issues they faced were just as pressing. This was certainly true of Chateau Woods near I-45 in southern Montgomery County.[33] Residents here struggled for two decades to refashion their subdivision into a town and, even more important to some, a community. After two decades of limping toward that end, residents soured on incorporation and elected a faction of residents whose central purpose was to deconstruct the legal entity of Chateau Woods systematically. Theirs was a creative destruction, they believed, clearing the way for the development's resurgence. The public and often very personal debates over disincorporation that ensued reveal the meaning these seemingly dry legal designations had for local residents. A battle raged for more than two years over whether Chateau Woods would be a town or a development, whether municipal government provided a community identity or perpetrated fiscal waste, and whether services should be provided by government or by business. The saga of Chateau Woods is a case study of the struggle between city and suburb.

Chateau Woods (originally Lake Chateau Woods) was one of the dozens of rural subdivisions that developers platted, advertised, and sold in Montgomery County during the mid-1960s. The construction of I-45 and TX-59, the opening of an international airport, and the continuing growth of Houston each pushed suburban sprawl north into Harris County and beyond. Integration and the looming specter of busing added an additional motive for suburbanites to settle across the county line in Montgomery

County. Lots in wooded nature only added to the appeal. Hughes Realty Company of Conroe and Texas State Land Associates sold the first lots in Chateau Woods in 1965, with all seven sections platted by the end of 1967. What homeowners in this 400-acre development received was standard for rural subdivisions—private gravel roads, septic tanks, electricity, and water on a seventy- by one-hundred-foot suburban lot.[34]

When the developer turned over responsibilities to homeowners some ten years later, Chateau Woods was already looking worse for wear. The country lanes that had seemed so picturesque and homey were now dotted with potholes and scarred with gullies to the point that the mail carrier and the school bus driver each thought twice before entering the development. Poor drainage made standing water a persistent nuisance. As for recreation facilities, they were almost nonexistent. The developer had left the Lake Chateau Woods Civic Club, formed in the late 1960s, with the responsibility for dealing with these pressing issues. A model of local participatory democracy on its face, the civic club's $14,000 was inadequate to the task of completing $60,000–$75,000 in drainage improvements and $350,000 in road improvements. The local county commissioner took over maintenance of the main parkway through the subdivision since it was a major thoroughfare, but he would not touch the other roads unless the residents brought them up to county standards first. The developer had cut corners to save money. His bankruptcy left the residents to bask in the privacy and exclusivity of their development as a tractor pulled their car out of a private ditch.[35]

With their troubles mounting, incorporation shone for many of these residents as a beacon of hope. A municipality would have the power to tax and could apply for federal and state grants. It would also protect them from any attempt by Houston to annex them. An October 1975 petition for incorporation received the signatures of one hundred registered voters, setting an election for December 6 of that year. In the ensuing months, incorporation opponents, most of whom were blue-collar workers or retirees, became increasingly vocal in their opposition. Warnings of increased taxation were the loudest, especially among those on fixed incomes. Some feared that incorporation would concentrate the power to tax "in the hands of a few councilmen" who would put their hands in their neighbors' pocketbooks. One resident, writing to the local newspaper, warned that "the initial charter may 'suggest' a small tax levy. But once the 'city' is declared legally an incorporation, and a council and other leaders chosen, the tax assessment might skyrocket 'surprisingly,' and dissidence can mean a court battle, which can be expensive and most probably futile." Such fears reflected the grassroots politics of much of the region. The same fear of urban control and annexation that drove many incorporations extended to the city that incorporation would create. A smaller, closer city held the same risks of political power and control. They feared that this new state structure would become a tool of oppression.[36]

When the votes were counted, incorporation carried the day 72 to 56. Only sixteen of the registered voters did not cast a ballot. The civic club and its allies won. A government consisting of a mayor and five aldermen was sworn in on April 17, 1976. With its newfound powers and income, Chateau Woods addressed some of the worst road problems and, perhaps more important, maintained the appearance of improvement. By 1980 these efforts had increased the population to 603 residents. The town also secured financing from the Texas Water Development Board (TWDB) to replace decaying and ineffective septic tanks with a sewer system.[37]

Over the next two decades Montgomery County continued to grow at a rapid pace. Chateau Woods stagnated. The town council struggled to provide the necessary services while relying on a tax base that included no commercial or industrial properties. The promised grants never seemed to materialize, and traffic fines and ordinance violations could do only so much to balance the budget. The services the town provided, therefore, were limited at best. The council managed to keep the roads teetering on the edge of passable. It contracted with an independent garbage service and maintained a volunteer dog catcher. The town police force consisted of eight to ten part-time, unpaid deputies, many of whom were Houston firemen. Their combined patrol hours covered only 15 percent of each week. In an emergency, the town relied on county units. For normal operations, residents got used to waiting. Such meager services may have been sufficient, or even quaint, in the 1970s. By the 1990s, however, the flourishing of The Woodlands, a master-planned community a few miles to the west, and the increasing professionalism of the Montgomery County government had significantly raised the expectations of potential homebuyers. Chateau Woods simply could not compete, leaving a large portion of the town in overgrown empty lots. Incorporation had failed to deliver.[38]

Confronted with grinding decay of the town's infrastructure, Mayor Reg Arceneaux, a former air force pilot and resident since 1985, moved to bring his constituents to the point of decision. On February 24, 1990, the mayor, acting in spite of opposition from the city council, sent to each resident a short document outlining the options available to the town. Residents would mark their preference knowing, as Arceneaux reminded them, that "any improvements to the City's road and/or drainage system will not be free." "It is up to you as individuals," he went on, "to best determine if the advantages of paved streets and improved drainage outweigh the increase in taxes required to pay for such improvements." Exactly how much they would have to pay remained unknown. Such estimates required engineering studies—studies they could not afford. Having warned his audience, Arceneaux proceeded. Option one was to trudge along as they had been, even though, as he explained, there was not enough money available "to maintain either the roads or drainage systems in a constant state of maintenance," let alone improve them. This option, he warned, would lead to the continued depreciation of their property values. The second option was to dissolve the town—to disincorporate and rejoin the county. This would allow their county commissioner to take over maintenance of their roads and ditches. Doing so would come at a cost. Not only could the county require them to bring their roads up to county standards first, they would also have to create a MUD to take over their water and sewer services and the debt they included. Eliminating the city would remove all their ordinances and police department and would leave property owners without deed restrictions or any other tools to prevent undesirable development. Arceneaux dismissed this option as overly complex, uncertain, and impractical. The mayor threw his weight behind a third option—taking on increased debt to fund repairs and thereby drive up property values and tax revenues to, in turn, service this debt. This increase would be substantial. Taking on another $1 million in debt would increase taxes from 70 cents to 100 cents per $100, increasing the average tax burden to $350–$500 per year. As residents went out to their mailboxes that winter evening, they received formal notice that they could either allow their town to continue to decay, literally wipe their town off the map, or significantly increase their tax burden. It was not good news.[39]

The returns from this mailing are not available, but its impact is clear. It sparked a movement for disincorporation that quickly gained momentum. Residents circulated a petition calling for a disincorporation election to be held on May 4, 1991. Blindsided by the movement, the mayor and town council scrambled to find a map to these seemingly uncharted waters. It was not that disincorporation was entirely unheard of. Between 1970 and 1977 there were 582 incorporations and eight-two disincorporations across the nation—a ratio of seven to one. Fourteen of these disincorporations occurred in Texas. One, Porter Heights Village, was only a dozen miles east of Chateau Woods.[40] Most of these, however, occurred soon after incorporation, as the burden of unpaid responsibilities and specialized knowledge required to run a city dawned on residents. These disincorporations were born of buyer's remorse. Chateau Woods residents, in contrast, were considering disincorporation only after their town was crumbling under their feet.

Disincorporation itself was relatively simple, so long as the municipality was not bound by debt obligations. It merely required a petition with signatures from two-thirds of registered voters and a majority vote on election day. For cities holding debt obligations, however, the logistics were more complicated. The city had to secure financing to cover its obligations and establish a legal entity to administer these payments in place of the municipality. If it failed to do so within forty-five days of disincorporation, the community would fall under a court-appointed receiver who had the authority to liquidate assets and raise taxes to settle the account. These were the facts city leaders learned when they contacted the state attorney's office and the TWDB. Armed with these facts, Mayor Arceneaux warned residents that an increasingly vocal block of residents was working to "impulsively and irresponsibly disincorporate our City on May 4th, regardless of the financial or long term consequences." Fortunately for Arceneaux, the petition failed on a technicality, in spite of receiving signatures from 86 percent of the registered voters. They would try again. Arceneaux was now unable to stifle the idea he had sparked.[41]

David C. and Lottie E. Schultz were at the forefront of disincorporation. These were hard-working, blue-collar, Texas conservatives—entrepreneurs and self-professed country people. David, whose roots in the area went back generations, owned and operated an excavation and dump truck business serving builders in The Woodlands. Lottie ran local beauty shops. The couple remembered moving to the development in October 1974, just as the movement for incorporation was gaining steam. The couple viewed incorporation as a mistake—a series of promises never fulfilled. As Lottie explained to a reporter in 1991, "We can't afford to play city anymore." "We don't have enough tax base to be a city. We've tried it for years and it hasn't worked."[42]

As David Schultz remembered it, he and three other men decided after reading Arceneaux's brief that disincorporation was the only real solution. They tried to convince the mayor to support their efforts. When he demurred, Schultz remembered saying, "Either you disincorporate, or we're going to kick each and every one of ya'll out of office and we're going to get into office, we're going to get elected, and we're going to do it. Because we want paved roads, clean ditches, and less taxes." Arceneaux, for his part, claimed that the disincorporationists promised him a barbecue and to "guarantee me re-election on May 4th" in exchange for his support. If he continued to oppose disincorporation, they would "guarantee my defeat." Whatever was actually said, the battle lines were drawn. Disincorporationists filed a second petition on March 18, 1991, containing enough signatures to force an election. Norma Caufield and Mattie Ola Parris counted signatures for the municipal government well into the evening. Two minutes before the stroke of midnight they reported to the mayor that the petition had received 223 signatures out of 359 qualified voters. It was seventeen short. The town, it seemed to Arceneaux and the city council, had once again narrowly averted disaster. Disincorporationists, in turn, interpreted this failure as evidence of an entrenched power structure that would do whatever it took to maintain control.[43]

With a municipal election looming, the disincorporationists selected John W. Brown, a retired businessman and resident since 1979, as a write-in candidate for mayor. They hosted their barbecue and did the legwork required to get him and two like-minded councilmen into office. With only two of the five council seats, however, there was only so much that the group could accomplish at that point. Brown's greatest achievement was refinancing the city's debt held by the TWDB. According to TWDB representative Bill Matthews, Brown achieved this $250,000 savings in exchange for a promise that the city would not disincorporate and that he would not support any movement to that end. This promise would hold until spring 1992, when both sides fought over three vacant council seats that would determine the fate of Chateau Woods.[44]

Historians have sketched the outlines of suburban politics at the national and metropolitan level: meritocratic individualism, freedom of association, taxation as fee-for-service, privatization, and a slavish commitment to maintaining property values against all perceived threats. These insights generally come from the analysis of the battles between suburbs and the metropolis; between these communities and external threats. Chateau Woods, in contrast, provides a glimpse into a self-contained suburban political fracas. Here residents fought over the proper role of local government, the value of community, the purpose of taxation, and who was qualified to lead. Their often shrill and petty debates reveal the way conservative political rhetoric played out at the subdivision level on the metropolitan fringe. For two years in the early 1990s, discord ruled Chateau Woods.[45]

The central issues at stake were relatively simple. The city's roads and drainage ditches were decaying rapidly. Of the city's income of $200,000, between $154,000 and $170,000 went to pay its bonded debt to the TWDB. This left $30,000–$46,000 to pay for the day-to-day operations of the city, including police protection, rent on the city hall building, and a small number of administrative employees. Whatever was left over after this was available for road maintenance. Given that the town had fewer than three hundred homes and no commercial or industrial tax base to speak of, raising additional funds through taxation was difficult. Legally the city could raise taxes only another fifty cents per $100 of appraised value, which would provide only somewhere in the neighborhood of $65,000 extra per year. Not only would this place a politically untenable burden on residents, it would also do little to address road and drainage problems, estimated to cost $1.5–$2.2 million and yearly maintenance estimated at $200,000. It was an untenable financial situation.[46]

Disincorporation, however, posed its own dangers. In addition to losing their local police, mayor, city council, and legal identity, residents would also lose the ordinances they relied on to protect their property values. These included everything from building codes and minimum home sizes to leash laws and health ordinances. Eliminating city hall would force the residents to rely on the county to provide these things. The one service the county government could not provide was deed restrictions. Local political leaders had rolled over all of these into city ordinances and had allowed the original restrictions to lapse. Forming new deed restrictions would require the approval of 75 percent of all the property owners. Given that only three hundred of the 1,100 platted lots had been built on, most of these were absentee owners, with many living out of state. Securing their approval would be costly and time consuming. Deed restrictions that did not include this undeveloped land, however, would be useless. The specter of imminent trailer park development on these lots became a favorite bugaboo summoned by the city's defenders. Each of these issues ultimately boiled down to the desire of homeowners to enjoy the suburban dream of quality of life, rising property values, and limited taxation. As one councilman put it, "I wan[t] to live in a nice place, not a dump." The question was whether this nice place would be best be secured within the boundaries of a city or on its legal ruins.[47]

Those who defended the town leveraged each of these issues to scare potential voters away from disincorporation. They did so as part of a larger defense of the public sphere—of the city as a means to transform residents pursuing individual self-interest into citizens pursuing the common good. For these town advocates, the persistence of Chateau Woods as an incorporated entity was vital to local control, local identity, and the common bonds of community. Debi Hall was the most prominent of these municipal defenders. Along with her husband Ronnie and her two children, Hall had moved to Chateau Woods in 1988. Tenacious and vocal to a fault, she quickly became a civic activist of the sort that drove the consumer movements of the 1940s and the grassroots environmentalism of the 1970s.[48]

Hall took charge of the struggling civic club, an entity one opponent labeled "a few acres of grass and a broke down swimming pool," and struggled to position it at the center of an engaged community. She failed. Local apathy and budget shortfalls forced her to donate the park to the city, abandon the club, and shut down the pool. This funneled all remaining hopes of community into the town. The civic club may have failed, but the town had picked up the pieces. When that was threatened, Hall and her allies raised a public outcry. They attacked at every point. Paving the roads would not only remove the rustic quaintness of the community, it would also put children at risk of being hit by cars. Property values would collapse if the town no longer controlled growth. Disincorporation would place their homes at the mercy of outside forces—whether this ended up being county officials, the nearby town of Oak Ridge North, or, even worse, the State of Texas. In each case, Hall continually turned the attention of residents back to the city as the anchor of their community. "Your home," she argued, "is your investment in this city. You must maintain your city as your home for the value to increase." Even as she ferociously attacked the disincorporationists, Hall continually called for a rejection of the "bitterness that has turned neighbor against neighbor." Above all, she clung to the dream of community self-reliance. "Patience and working together," she argued, "can accomplish a lot more than depending on someone else to do it for us." Hall and her allies fought to save the dream of small-town America within a crumbling rural subdivision.[49]

Disincorporationists crafted their own political narrative to undermine support for the city. Drawing on Reagan-era rhetoric, these small-business leaders crammed the amateurish local government into the mold of an entrenched, scheming, corrupt local bureaucracy. Here, Schultz and his cohorts believed, was a local head of the hydra of big government that had slithered its way into every nook and cranny of American life. Looking back on these events, Schultz remembered:

> Naturally they didn't want to disincorporate, because the city secretary got a check, the city marshal got a check, all of them got checks, you know small checks, but they got checks. But the main thing was they had power. They could come out here and say, like that truck parked out here on the shoulder road, the cop could stop here and say, 'you can't park there on the shoulder road over ten minutes, city ordinance,' and people doesn't realize it, but a city councilman or mayor has as much power as the President of the United States. As long as it is in this subdivision. I can make your life a living Hell in here if I'm a city councilmen and the rest of them agree with me. Because you can pass ordinances, the same thing as a law.[50]

These local officials, according to Schultz, were petty despots who cloaked their "power tripping" behind the veil of community. That is not to say Schultz was categorically opposed to government per se. Rather, he opposed taxation, spending, and most regulation. It was the perceived inefficiencies, incompetence, and wastefulness of local government that drove Schultz into local politics to begin with. In contrast with Hall, Schultz styled himself a grassroots populist figure, a "country boy" leader of "hillbillies and hicks and rednecks" who "said what needed to be said in 'plain Jane' words, didn't pull no punches" and worked to "make this a better place to live." Government abuses forced him into politics. Once disincorporation was accomplished, he would ride into the sunset. As he later remembered, "I don't want nothing to do with no politics."[51]

As these conflicts became increasingly personal and vitriolic, they also became explicitly gendered. Disincorporationists were led, in Schultz's words, by six men "and their wives." Although they would eventually elect one councilwoman, thirty-year-old teacher Kerry Mathes, in 1991, the face of disincorporation was that of a businessman. These men had, they

believed, left the city of Chateau Woods in women's hands, and the result
was a struggling, pathetic town. These city leaders had naively called for
community as they ineptly squandered local tax money. Now that things
were getting serious, it was time for experienced men to step in address
a situation where, in the words of one letter to the editor, "we have some
women that are on a ego trip and want to hold office." Reg Arceneaux,
responding to a political statement from the Hall camp, claimed, "My
6th grader could not believe the spelling and grammatical errors in your
letter." Arceneaux went on to verbally shove Hall back into the women's
sphere of "community" building.[52]

Hall's political actions—turning to half-truths and fear-mongering,
verbally haranguing her opponents in public, and allegedly vandalizing
opponents' political signs—certainly gave her opponents plenty to complain about. Disincorporationists used these opportunities to emphasize
their business credentials and claim the mantle of dispassionate rationality.
They spoke of the city not as a focal point for community but as an ineffi-
cient and ineffective provider of services. As former mayor and eventual
disincorporation supporter Reg Arceneaux argued in response to a Debbi
Hall missive, "Our city exists only in terms of legal definition, not in terms
of capability to provide services to its residents." Cutting costs, improving
services, cutting taxes, and renegotiating debt were their central goals. The
disincorporationists cared far less about relations among residents than
they did about the relationships between the subdivision and the county,
service providers, and the State of Texas.[53]

Debbi Hall, Ida Ditto, and Betty Maduzia blanketed the city with election pamphlets. One letter claimed that the Justice Department in Washington, the state attorney general's office, and the state election division
were "investigating" the disincorporationists. Tragically, the rancorous
election shattered the very community these city officials claimed to
defend. It turned neighbor against neighbor and sowed the seeds of bitterness deep in multiple generations. Frustrated by the discord and tired of
having to confront his friends and neighbors, the city police chief resigned
on March 26. At the May 2 election voters decisively rejected Hall, Aycock,
and Hagan and elected the disincorporationists: Bale, Schultz, and Simpson. Disincorporationists now controlled the city government.[54]

The disincorporationists ran their first city council meeting on May
14 with the decisiveness of a third-world coup. They eviscerated the government of Chateau Woods. As Schultz remembered, "We either fired or

accepted resignations from every person that had anything to do with the city government. The first night." This included Ida Ditto (city comptroller) and Betty Maduzia (municipal judge), four police officers, the police chief, the planning commission, and the dog catcher. They negotiated their way out of a lease on the city hall along the highway and returned the government to the civic building. Their opponents cried foul, claiming this was a "takeover by power-hungry people." Secretary Mary Cleere withheld city documents. Ida Ditto refused to turn over city financial records until she could have them audited, to protect herself against possible prosecution for mismanaging city funds. Even Mayor Brown was aghast at the fallout, telling reporters, "It's so divided it's unreal." Debi Hall, for her part, began videotaping the city council meetings under the authority of the state open meetings act. The council, in response, forced her to do so from a small, dimly lit corner of the room behind a taped line on the floor. When on one occasion she crossed that line, they summarily had the police chief escort her from the meeting. With all this chaos going on within city hall, residents became increasingly confident that the city was not long for this world and began flagrantly disregarding city ordinances.[55]

By August 1992 the new anti-city city officials had reduced the tax rate from ninety-two to eighty-eight cents per $100, renegotiated rates with the telephone company, and hired a city attorney to protect them from further political opposition. Disincorporation, however, would have to wait until the next election, on May 1, 1993. The councilmen secured a written agreement from county commissioner Ed Chance that he would accept the roads without condition once they disincorporated. They also began the process of preparing a MUD to take control of the city's debt service once the city dissolved. The problem was that creating a MUD could cost anywhere between $5,000 and $30,000 in legal and government fees. That was money Chateau Woods simply did not have. Their proposed solution was to have Bob Rabuck, their state representative, shepherd a bill through the state legislature that would create a Chateau Woods MUD with special provisions giving it the power to enforce building ordinances. This became the focus of their efforts and a beacon of hope for other struggling municipalities. The legislature, however, had little incentive to follow the city council's timeline.[56]

Finally, in January 1993, five months before the election, word came back that the Texas constitution did not allow the state legislature to disincorporate a city or create a MUD. Only voters could do these things. A bill creating the MUD and disincorporating the city—with voter consent for each—made it out of committee. The provision allowing the MUD to enforce codes and ordinances, however, did not. The state was unwilling to allow the troubles of one town to create a precedent for such a large extension of MUD powers. Opponents seized on this to make one final stand against disincorporation in the weeks leading up to the May 1 election. It was too late. A March petition received 339 out of 434 voters, putting disincorporation on the ballot. The county commissioners also passed a motion in support of the law. When election day arrived, voters approved disincorporation 233 to 90. Mayor Brown and his fellow incumbents won their seats by similar margins. Unfortunately, the law had not yet cleared the senate and therefore the MUD was not on the ballot. The governor ultimately signed the law on June 16. This left disincorporation caught in legal limbo until the town held its final election, on August 14, 1993, with ample time to spare before the October bond payment was due. The final vote was 205 to 51. The city of Chateau Woods was no more.[57]

This was not the end, however, for John Brown and Greg Simpson's political careers. Each transitioned into leadership of the MUD board—a position where each could exercise power over the community's finances without the public hassle of municipal politics. Disincorporationists successfully filed deed restrictions on April 25, 1994, resolving this final legal hurdle. When the county commissioner asphalted the roads in 1996, the community was finally open for new growth. From that point, the development filled out, property values rose, and Chateau Woods once again offered newcomers a picture-perfect suburban lifestyle.[58]

The story of Chateau Woods is many things. For Schultz, Brown, and their colleagues, it was a populist romance—a classic tale of everyday heroes struggling against the odds to defeat the city and its entrenched elites. For Hall and her allies, it was a tragedy—a tale of crass business opportunism crushing an authentic community and destroying a struggling town and replacing it with a privatized subdivision. Seen through a wider lens, neither reading is particularly satisfying. Rather, the story of Chateau Woods serves as a case study in the perils of hyper-local, amateur-led democracy. These residents left the city, then constructed their own city to prevent annexation by Houston and provide local services. The new city then proceeded to fail in ways that far outpaced anything Houston could have achieved. In this the residents got more than they bargained for. Creating and destroying a city channeled the political, emotional, personal, and financial energies of this small subdivision inward, creating a frothing mess of backbiting and strife. This is not to say that local democracy could not have worked. It does suggest, though, that the contentious national political rhetoric of the period bore some of its most bitter fruit within local communities. Chateau Woods was too small, too ingrown, too narrow in its interests to sustain a healthy democracy. It was only when the subdivision rejoined the more professionalized and politically stable county government that Chateau Woods stopped tearing itself apart.

Taken more broadly, Chateau Woods also suggests that incorporation was not always an answer to suburban problems on the metropolitan fringe. Historians should be careful about assuming that white privilege and suburban secession guaranteed success. Rather, they need to make room in their narratives for those suburbs that did not succeed. If this is the crabgrass frontier, where is the suburban Donner party? Where are the plywood signs promising brick split-levels set in beautiful nature, rotted and swaying in the breeze—headstones for empty plats and unfulfilled dreams slowly disappearing behind poison ivy and Virginia creeper? Disincorporation may have saved Chateau Woods from such a fate, but at great cost.

NOTES

1. Dolores Hayden, *Building Suburbia: Green Fields and Urban Growth, 1820-2000* (New York: Vintage Books, 2003, 2004); Kevin M. Kruse and Thomas J. Sugrue, eds., *The New Suburban History* (Chicago: University of Chicago Press), 2006.

2. Stanley K. Schultz, *Constructing Urban Culture: American Cities and City Planning, 1800-1920* (Philadelphia: Temple University Press, 1989), 66–75; Robert D. Thomas and Richard W. Murray, *Progrowth Politics: Change and Governance in Houston* (Berkeley, CA: Institute of Governmental Studies and International and Area Studies, 1991), 99–112; J. C. Davis Jr., "Legal Aspects of the County Commissioners Court," speech before the County Judges and Commissioners Conference, College Station, TX, February 25, 1959, in Folder Speeches by J. C. Davis Jr., Box 2004/082–1, Attorney General Records, Texas State Library, Austin; Commissioners Courts—Subdivision Plats Filing and Recording in Certain Counties, H.B. No. 289, Reg. Sess., 55th Leg., (1957), 1302–3.

3. Evan McKenzie, *Privatopia: Homeowner Associations and the Rise of Residential Private Government* (New Haven, CT: Yale University Press, 1994), 11; Joel Garreau, *Edge City: Life on the New Frontier* (New York: Doubleday, 1991), 179–208.

4. Municipal Utility Districts, H.B. No. 1458, Reg. Sess., 62nd Leg., (1971), 1.774–812; Virginia Marion Perrenod, *Special Districts, Special Purposes: Fringe Governments and Urban Problems in the Houston Area* (College Station: Texas A&M University Press, 1984), 17–18, 34. A 1978 amendment extended their powers to cover firefighting as well.

5. May 1971 and April 1977, Montgomery County Commissioners' Court Records, Montgomery County Annex, Conroe, TX.

6. H.B. No. 1458 (1971); Thomas and Murray, *Progrowth Politics,* 117–35. If the number of property owners was over fifty, then at least fifty had to petition for the MUD to be considered by the state.

7. Perrenod, *Special Districts,* 117. For a counterargument, see Robert B. Hawkins Jr., *Self Government by District: Myth and Reality* (Stanford, CA: Hoover Institution Press, 1976).

8. Kenneth J. Jackson, *Crabgrass Frontier: The Suburbanization of the United States* (New York: Oxford University Press, 1985), 138–56; Robert O. Self, *American Babylon: Race and the Struggle for Postwar Oakland* (Princeton, NJ: Princeton University Press), 2005; Colin Gordon, *Mapping Decline: St. Louis and the Fate of the American City* (Philadelphia: University of Pennsylvania Press, 2008).

9. U.S. Bureau of the Census, *Boundary and Annexation Survey, 1970–1977,* GE 30–3 (Washington, DC: U.S. Government Printing Office, 1979), 2.

10. Mary M. Edwards, "Understanding the Complexities of Annexation," *Journal of Planning Literature* 23 (November 2008): 119.

11. Cities and Towns—Authorizes Cities of More than 5000 Inhabitants to Adopt and Amend Their Charters, H.B. No. 13, Regular Session, 33rd Leg., 307–18; House Research Organization, *Annexation Vexation,* Texas House of Representatives, 75–3 (1997). General Law Municipalities (under 5,000 residents) require the consent of the majority of the voters in an area to annex. The power of home rule cities to annex unilaterally was upheld by the courts in *Cohen v. City of Houston,* 176 SW 809; and *Dallas*

County Water Control and Improvement District No. 3 v. City of Dallas, 149 Tex. 362 and 233SW 2d 291 (1950); Texas Legislative Council, *Municipal Annexation*, Texas Legislature, 56–6 (1960). See also David Brooks, *Texas Practice*, 2nd ed., vol. 22: *Municipal Law and Practice* (St. Paul, MN: West, 2000), 66–89.

12. Texas Legislative Council, *Municipal Annexation*, 5–6, 32; Eddie Hughes, "City Completely Hemmed In by 27 Other County Towns," *Dallas Morning News*, August 21, 1960, 15; Brooks, *Texas Practice*, vol. 22, 84–85, and vol. 23A, app. G, Article 1, §2b.

13. Carl Abbott, *The New Urban America* (rev. ed., Chapel Hill: University of North Carolina Press, 1987), 54–55; Char Miller and Heywood T. Sanders, "Olmos Park and the Creation of a Suburban Bastion, 1927–1939," in Miller and Sanders, eds., *Urban Texas: Politics and Development* (College Station: Texas A&M University Press, 1990), 113–27; Letitia A. Gomez, "Growth Management and Annexation: San Antonio's Struggle with Growth," Master's thesis, Trinity University, 1987; Arnold Fleischmann, "Sunbelt Boosterism: The Politics of Postwar Growth and Annexation in San Antonio," in David C. Perry and Alfred J. Watkins, eds., *The Rise of the Sunbelt Cities* (Beverly Hills, CA: Sage, 1977).

14. Kyle Shelton, "Houston (Un)limited: Path-Dependent Annexation and Highway Practices in an American Metropolis," *Transfers* 4 (Spring 2014): 97–115; Thomas and Murray, *Progrowth Politics*, 141–77.

15. The city further annexed a strip of land around Pasadena in 1949, preventing it from additional expansion. David G. McComb, *Houston: The Bayou City* (Austin: University of Texas Press, 1969), 200–205.

16. Texas Legislative Council, *Municipal Annexation*, 47, 51; McComb, *Houston*, 200–205; Louie Welch, untitled speech, American Municipal Association Convention, Boston, MA, folder 7, box 1, Lewis and Catherine Cutrer Collection, Houston Metropolitan Research Center, Houston Public Library, Houston, TX (hereafter Cutrer Collection).

17. Abbott, *New Urban America*, 268.

18. Texas Legislative Council, *Municipal Annexation*, 33, 59.

19. "Plano and Irving Rush Annexations," *Dallas Morning News*, August 13, 1960, 2; Texas Legislative Council, *Municipal Annexation*; McComb, *Houston*, 200–205. "Houston Annexes All of County Left," *Houston Post*, June 23, 1960; "Annexations Pass Silly Stage; Houston Must Protect Future," *Houston Chronicle*, June 20, 1960 (quotation) both in E. A. Lyons Jr. Scrapbooks, Harris County Archives, Houston, TX.

20. "Plano and Irving Rush Annexations," 2; Texas Legislative Council, *Municipal Annexation*, 38–40, 55; "Wise Tax Policy on Annexed Land," *Dallas Morning News*, July 20, 1960, 2; Hughes, "City Completely Hemmed In," 15.

21. Eugene Maier, Director of Public Works to Mayor Lewis Cutrer, April 27, 1961, folder 4, and Ralph S. Ellifrit, Director of City Planning, "Allocation of Small Cities in a Metropolitan Area," April 27, 1961, folder 7, both in box 1, Cutrer Collection.

22. Hughes, "City Completely Hemmed In," 15 (quote).

23. Municipal Annexation Act, H.B. 13, 3rd Sess., 57th Leg. (1962), 447–54; Brooks, *Municipal Law and Practice*, vol. 22, 73–77; and Thomas and Murray, *Progrowth Politics*, 155–59.

24. In terms of land area annexed during 1970–77, Houston, San Antonio, Dallas, Corpus Christi, El Paso, and Austin were ninth, tenth, eleventh, twelfth, fifteenth, and

twenty-second in the nation, respectively. If Alaskan cities are removed, those numbers become fifth, sixth, seventh, eighth, eleventh, and eighteenth. U.S. Bureau of the Census, *Boundary and Annexation Survey, 1970–1977*, 18, 72.

25. "Houston Annexation Move Cuts into Montgomery Co.," *Conroe Courier*, August 22, 1965, 1; Tom Bacon, "County, City of Houston Conflict Likely to Grow," *Conroe Courier*, December 16, 1970, sec. 1, 2.

26. "Big Brother's Cut off," *Houston Chronicle*, June 15, 1969, 30; "Makes First Annexation Move," *Conroe Courier*, March 9, 1966, 1; "Hooray for Conroe's 'Finger-Annexation'!," *Conroe Courier*, December 11, 1966, sec. 2, 11. The legislature acted to limit this tactic when it increased the minimum width of land to 500 feet in 1977 and 1,000 ft in 1987. House Research Organization, *Annexation Vexation*; An Act Relating to Annexation Authority of Municipalities, S.B. 962, Reg. Sess., 70th Leg. (1987), sec. 3, 3674–83.

27. Kathryn T. Rice, Leora S. Waldner, and Russell M. Smith, "Why New Cities Form: An Examination into Municipal Incorporation in the United States 1950–2010," *Journal of Planning Literature* 29, no. 2 (2014): 147; Agustin Leon-Moreta, "Municipal Incorporation in the United States," *Urban Studies* 52, no. 16 (2015): 3162; Abbott, *New Urban America*, 182.

28. Texas Legislative Council, *Municipal Annexation*, 15; Abbott, *New Urban America*, 54–55, 177.

29. U.S. Census Bureau, *Boundary and Annexation Survey*, tab. 9; "Houston Annexation Move Cuts into Montgomery Co.," 1.

30. Oak Ridge North, Whispering Oaks, Shenandoah, Chateau Woods, Woodbranch Village, Panorama Village, Porter Heights, Chapel Lakes, Splendora, Patton Village, and Stagecoach Farms all incorporated during this period. Three other developments had their incorporation attempts voted down. For records of incorporation elections, see August 1966, December 1966, April 1969, June 1972, May 1973, June 1973, October 1973, March 1974, June 1974, December 1975, August 1977, and February 1979, all in Montgomery County Commissioners' Court Records, Montgomery County Annex, Conroe, TX; and April 7, 1980, vol. 30, 745, County Commissioner Minutes, Montgomery County Annex, Conroe, TX. See also "New Towns Popular Now," *Conroe Courier*, May 26, 1974, 8.

31. Mary Goranson Eklof, "City of Oak Ridge North," in History Book Committee, Montgomery County Historical Society, *Montgomery County History, 1981* (Winston-Salem, NC: Hunter, 1980), 54; "Historical Background & Environmental Setting, Oak Ridge North, Texas," Folder City History, Oak Ridge North Municipal Records, Oak Ridge North, TX.

32. Andrew C. Baker, *Bulldozer Revolutions: A Rural History of the Metropolitan South* (Athens: University of Georgia Press, 2018), chap. 4.

33. Along with the sources cited individually, the following extended discussion of Chateau Woods is informed by my interview with David C. Schultz, August 8, 2012, Chateau Woods, TX, in author's possession; and by documents found in the David C. Schultz personal collection, Chateau Woods (hereafter Schultz Collection).

34. Deed Books, vol. 589, 459, 461, 463, 465, Montgomery County Clerk Online Records (hereafter MCCOR), https://govapps1.propertyinfo.com/tx-montgomery (accessed May 1, 2015).

35. "Lake Chateau Woods Incorporation Sought," *Conroe Courier*, October 21, 1975, 1; "The History of Chateau Woods," in History Book Committee, *Montgomery County History;* "Incorporation Could Finance Repairing Roads, Drainage," *Conroe Courier,* November 2, 1975.

36. "County Court to Get Incorporation Petition," *Conroe Courier*, October 27, 1975; "Incorporation Election Is Scheduled Dec. 6," *Conroe Courier*, October 28, 1975; "Opposition Forming to Incorporation," *Conroe Daily Courier*, October 29, [1975], (first quotation); Connie Baker, "Lake Chateau Woods People: Look at Incorporation Again," *Conroe Daily Courier*, October 31, [1975] (second quotation).

37. "History of Chateau Woods."

38. Reg Arceneaux, To the Citizens of Chateau Woods, January 10, 1992, Schultz Collection.

39. The city was in such poor financial straits that the mayor was unwilling to pay for an election unless he was sure a measure would pass. Reg Arceneaux, City of Chateau Woods Preference Poll, Road and Drainage Improvements, February 24, 1990, Schultz Collection (quotations), Tax rate taken from Reg Arceneaux to the Residents of Chateau Woods, April 24, 1991, Schultz Collection.

40. U.S. Census Bureau, *Boundary and Annexation Survey*, 1, 86; Tim Cumings, "Change Due for Porter Heights?," *Conroe Courier*, February 9, 1973, 1, 8. See also "2d De-incorporation Election Requested," *Houston Chronicle*, May 17, 1955.

41. Reg Arceneaux to the Residents of Chateau Woods, April 24, 1991, Schultz Collection; Arceneaux, To the Citizens of Chateau Woods, January 10, 1992, Schultz Collection (quote). The lawyer drawing it up used the plat name Lake Chateau Woods rather than the town name Chateau Woods.

42. The couple purchased two lots in the town in 1976. Deed Books, vol. 935, 75, MCCOR; Paul McKay, "A City at Its Limits," *Houston Chronicle*, June 30, 1991, 1C, 3C (quotations).

43. Schultz did not identify the three other men. Later records indicate that they were likely Greg Simpson, Larry Bell, and John Brown; Reg Arceneaux to the Residents of Chateau Woods, April 24, 1991, 3; Reg Arceneaux to Norma B. Caufield and Mattie Ola Parrish, March 18, 1991, and Norma B. Caufield and Mattie Ola Parrish to Reg Arceneaux, March 18, 1991, both in Shultz Collection; McKay, "City at Its Limits," 1C, 3C.

44. Kelly Schlesinger, "Chateau Woods Write-in Votes Elect 3 Candidates," *Conroe Courier* [undated], Schultz Collection; Lori Gula, "Efforts to Get Chateau Woods Streets Paved Raise Issues," *Conroe Courier*, April 26, 1992, 1-A, 2-A; John Brown, memo, "To the Residents of Chateau Woods" [October 1992], Schultz Collection. I have been unable to confirm this promise beyond newspaper coverage.

45. Kevin M. Kruse, *White Flight: Atlanta and the Making of Modern Conservatism* (Princeton, NJ: Princeton University Press, 2005); Matthew D. Lassiter, *The Silent Majority: Suburban Politics in the Sunbelt South* (Princeton, NJ: Princeton University Press, 2006); Kruse and Sugrue, *New Suburban History.*

46. Ed Chance to John Brown, March 27, 1992, Schultz Collection; Gula, "Efforts to Get Chateau Woods Streets Paved," 1-A, 2-A; Larry Bale to Robert Harry Rabuck, July 22, 1992, Schultz Collection.

47. Larry Bale to Robert Harry Rabuck, July 22, 1992, Schultz Collection; Greg

Kubeska, "Let's All Work Together to Clean Our City Up," January 9, 1992, Schultz Collection (quotation).

48. File 8815391, MCCOR; Lizabeth Cohen, *A Consumers' Republic: The Politics of Mass Consumption in Postwar America* (New York: Knopf, 2003); Adam Rome, *The Genius of Earth Day: How a 1970 Teach-In Unexpectedly Made the First Green Generation* (New York: Hill and Wang, 2013); Lois Marie Gibbs, *Love Canal and the Birth of the Environmental Health Movement* (Washington, DC: Island Press, 2011).

49. Reg Arceneaux to Fellow Residents of Chateau Woods, April 2, 1992 (first quotation); Debi Hall, "To the Citizens of Chateau Woods," March 21, 1992 (fourth quotation); Debi Hall, "To the Residents of Chateau Woods," March 29, 1992 (fifth and sixth quotations); Reg Arceneaux, "A Public Note to Mrs. Hall, April 28, 1992; Committee to Elect Debi Hall, Political Advertisement, April 4, 1992 (second and third quotations), all in Schultz Collection.

50. Schultz interview.

51. Schultz interview (quotations); Kirbow, Osburn, and Owen, "To the Residents of Chateau Woods," April 10, 1993, Schultz Collection.

52. Reg Arceneaux, "A Public Note to Mrs. Hall," April 28, 1992, Schultz Collection.

53. Kelly Schlesinger, "Chateau Woods Write-In Votes Elect 3 Candidates," *Conroe Courier* [undated]; Helen Wimhish, letter to the editor [undated, likely *Conroe Courier* but lacking masthead]; Arceneaux, "A Public Note to Mrs. Hall," April 28, 1992 (third quotation); political advertisement, Committees to Elect Larry Bale, David Schultz, and Greg Simpson, April 13, 1992; John Brown, "To the Friends and Neighbors of Chateau Woods," [undated]; Arceneaux to Residents, April 2, 1992 (fifth quotation); John Brown, memo, "To the Residents of Chateau Woods," [October 1992], all in Schultz Collection.

54. Ditto, Hall, and Maduzia, "An Urgent Message" [April 1992] (quotation); Gary W. Miller to John Brown and City Council, March 26, 1992; Kerry L. Brusky, "Chateau Woods Mulls City's Disincorporation," *Houston Chronicle*, June 3, 1992; "New Council Off Base," newspaper clipping [undated]; Paul McKay, "Chateau Woods Official Fired amid Meeting Issue," *Houston Chronicle*, May 19, 1992, 15A, all in Schultz Collection.

55. City Council Minutes, May 14, 1992; McKay, "Chateau Woods Official Fired," 15A (second and third quotations); John Brown to Mary Cleere, May 7, 1992; Tim Wesselman, "Former Chateau Woods City Treasurer Demands Audit," *Conroe Courier*, June 12, 1992, 1-A, 2-A; Tim Wesselman, "Chateau Woods Nears Decision on Its Future," *Conroe Courier* [undated]; Suzanne West, "City Looks at Disincorporation" [newspaper clipping], August 19, 1992; Kenneth R. Mullins to Mayor and City Council, May 1, 1992, all in Schultz Collection.

56. City of Chateau Woods, "Newsletter" [August 1992]; Michael C. Beller to Bill Corsbie, October 7, 1992; Gula, "Efforts to Get Chateau Woods Streets Paved Raise Issues," 1-A, 2-A; Larry Bale to Robert Harry Rabuck, July 22, 1992; West, "City Looks at Disincorporation"; Bill Corsbie to Ms. Sue Loredo c/o Bob Rabuck, October 9, 1992; Bill Corslie to Bob Rabuck, October 19, 1992, all in Schultz Collection.

57. Bob Rabuck to John Brown, January 12, 1993; Betty Maduzia, Debi Hall, Harry Turley, political flier, 1993; Ida B. Ditto, to Residents of Chateau Woods, April 27, 1993; Kirbow, Osburn, and Owen, "To the Residents of Chateau Woods," April 10, 1993; Tim Wesselman, "End of City Government Voters' Wish" [newspaper clipping], March 5,

1993, 1-A, 5-A; Tim Wesselman, "Chateau Woods Voters Favor Disincorporation," *Conroe Courier*, May 1993; Lori Gula, "New Controversy Emerges as Chateau Woods Vote at Hand," *Conroe Courier*, August 14, 1993, 1-A, 6-A; Kerry Brusky, "Chateau Woods May Create MUD District," *Houston Chronicle*, May 5, 1993, 8; Resolution Declaring the Votes of the Election Abolishing the City of Chateau Woods . . . [1993], all in Schultz Collection. An act relating to the creation . . . of the Chateau Woods Municipal Utility District and the abolition of the city of Chateau Woods, H.B. 2815, Reg. Sess., 73rd Leg., sec. 2, 2908–13; Montgomery County, Texas, Commissioners Court Minutes, March 22, 1993, www.mctx.org/cclerk/ccminutes (accessed May 1, 2016).

58. Resolution Declaring the Votes, Schultz Collection; Lori Gula, "New Controversy Emerges as Chateau Woods Vote at Hand," *Conroe Courier*, August 14, 1993, 1-A, 6-A; File 9422548, MCCOR; Montgomery County, Texas, Commissioners Court Minutes, June 24, July 15, October 7, 1996. Larry and Lottie Schultz looked to invest in this new growth, purchasing three lots in the subdivision between October 1991 and December 1993. They split the third with fellow disincorporationists Connie and Larry Bale. Disincorporation, for them, was always about increasing property values. It is no surprise, then, that they bet on the financial benefits of their political activities. Files 9147557 and 9364378, MCCOR.

{ 4 }

SUBURBAN MIGRATION PATTERNS OF AFRICAN AMERICANS IN DALLAS

THEODORE M. LAWE AND GWENDOLYN M. LAWE

Considerable African American emigration out of Dallas, Texas, to surrounding suburbs started in the 1950s with the opening of a planned new community in Hamilton Park and the simultaneous development of an all African American professional housing development. African Americans moved from older neighborhoods to South Dallas, Oak Cliff, and suburban communities north and south of Dallas. Rising incomes, civil rights legislation, Supreme Court decisions, and active local African American leadership aided this "black flight" to Texas suburbia.

With a population of over one million, Dallas is the second-largest city in Texas and the eighth-largest city in the United States. African Americans are approximately 25 percent of the population, Hispanics approximately 30 percent, and whites 45 percent. Historically Dallas had a majority white population until 2000. Because of the "white flight" era of white residential movement to suburbia in the 1970s and 1980s, however, the growth in the Dallas population has come from Hispanics, African Americans, and other minorities.

The aims of this chapter are to tell the story of African American suburban movements within the context of African American history in Dallas; to highlight some of the differences in African American migration patterns between northern, eastern, and southern suburbs; and to comment on the negative effect suburban migration is seemingly having on Dallas and its ability to stay competitive with its surrounding suburbs.

The Dallas metropolitan area has over forty incorporated suburban communities, and approximately half of them are located in Dallas County. The other half are located in Collin and Denton counties.[1] For the purposes

of this chapter, suburban communities located in Dallas County are considered "inner-suburban communities," and suburban communities outside of Dallas County are considered "outer-suburban communities." The
inner-suburban communities are the oldest and best established. To the
north of Dallas inner-suburban communities include Richardson, Farmers
Branch, Carrollton, and Coppell. The newer outer-suburban communities
north of Dallas are Plano, Allen, Frisco, Flower Mound, Prosper, and Mur-
phy. Several of these are still growing.

The primary suburbs to the east of Dallas are Mesquite, Sunnyvale,
and Rockwall. To the south of Dallas with shared boundaries are such
inner-suburban communities as Duncanville, DeSoto, Glen Heights, and
Lancaster. Cedar Hill is the only suburban community without a common
boundary with Dallas. To the west of Dallas are several cities that are fully
developed such as Irving, Grand Prairie, and Arlington. Because they are
not entirely as dependent upon Dallas, as well as being covered more thoroughly by Robert B. Fairbanks and Philip G. Pope in their contributions
to this volume, these suburb—or "suburban cities"—are not a part of this
analysis.

African American movement to the suburban communities has been
a function of both push and pull—flight and exodus. Similar to older
narratives concerning the suburban dream, including historian Andrew
Wiese's authoritative account of African American suburbanization,
African Americans have been pushed out of overcrowded and dilapidated
neighborhoods and lured to suburbia in pursuit of improved quality of life
and social status.[2] From the end of the Civil War through the early twentieth century, African Americans were confined to certain parts of the city
because of Jim Crow laws and customs. In 1916 the City of Dallas officially
passed the first of several restrictive resolutions and ordinances that dictated where African Americans could live; the National Association for the
Advancement of Colored People (NAACP) played a vital role in litigating
several residential segregation cases. Perhaps unsurprisingly, the African
American community was unstable, with all social, educational, and economic classes of people living together.[3]

It was not until after World War I that African Americans in Dallas
openly decided that they would foster community within the confines of
Jim Crow. To address race relation issues, for instance, they organized one
of the first chapters of the NAACP in Texas in 1918; the coordination of
civic affairs fell to the Dallas Negro Chamber of Commerce, established

in 1926 (the name changed to the Dallas Black Chamber of Commerce in 1975). By the 1930s, African Americans had elected a "bronze mayor" to symphonically represent their interests, with both Edgar E. Ward and A. A. Braswell holding the position (Ronald Kirk became the first official African American mayor of Dallas, serving from 1995 to 2002). In addition, by the 1940s and beyond, African Americans challenged racist local power structures and made inroads into Dallas area politics. For example, though unsuccessful, Maynard H. Jackson, one of the leaders in the Progressive Voters League, became the first African American to run for the Dallas school board; in 1959, attorney C. B. Bunkley also ran, but lost.[4]

A turning point for African Americans in Dallas, however, came earlier, in 1936, when Dallas's African American leadership, through the Dallas Black Chamber of Commerce, secured a $100,000 federal grant to host an exhibition at the Centennial Texas State Fair. The exhibition was called the "Hall of Negro Life." It showcased African American achievements and allowed Dallas African American leaders and others to actively plan and set goals for the community and other institutional changes throughout the state of Texas. The plans audaciously called for overturning the Democratic white primary (which occurred with *Smith v. Allwright*, 1944), adjusting salaries for public school teachers (Gilmer-Aikin Law of 1949), and integrating the University of Texas Law School (*Sweatt v. Painter*, 1950). All of these goals were achieved with the assistance of the Legal Defense Fund of the NAACP and federal and state courts. In fact, Dallas was the first major city to equalize pay for teachers in the state of Texas (1943).[5]

Political progress also began to pay bigger dividends in the 1940s when an African American became a precinct chair in the Democratic Party and participated in the Democratic county convention for the first time. During this time the Dallas city council also authorized the hiring of fourteen African American police officers with limited authority to patrol the streets in African American communities. The city council also authorized the building of several segregated public housing projects in West Dallas for African Americans and a middle-class housing subdivision to maintain residential segregation in an area that became known as Hamilton Park, named for Dr. R. T. Hamilton, an influential African American physician.[6]

Although Hamilton Park deserves a lot of our attention, other residential communities for African Americans in Dallas should be noted as well, including Deep Ellum (East Dallas), Joppa (South Dallas), Ten-Mile Creek (South Dallas), Elm Thicket/North Park (northwestern Dallas), Little Egypt (North Dallas), and Freedmen Town (North Dallas). The largest of these settlements was Freedmen Town, located north of downtown near the Central Expressway.[7] This area was demolished in the late 1990s and rebuilt as a swanky gentrified neighborhood of townhouses and apartments, with less than 1 percent of the residents being African American (today the area is known as Uptown).

As historian William H. Wilson has authoritatively chronicled, Hamilton Park was a planned African American suburban community in the 1950s. It was the first attempt for middle-class African American families to own new housing versus the so-called leftover approach whereby they moved into vacated, older white neighborhoods. In their attempt to maintain segregation, Hamilton Park was founded, funded, and supported by the white leadership in Dallas. The famed Hoblitzelle Foundation of Dallas, for instance, made loans of $217,000 to acquire the land, and the three leading banks in Dallas made loans of $424,000 to fund the utilities and subdivision regulations. The Federal Housing Authority (FHA) and Veterans Affairs, never ones to aid African American suburbanization actively but rather ensure residential segregation in America, assisted with the financing of the mortgages.[8]

Organizing for the development of the Hamilton Park subdivision started in the mid-1940s when Dallas had an African American population of slightly more than 50,000. The City of Dallas and the "business elite" organization known as the Dallas Citizens Council hired the land-use planning firm of Harland Bartholomew. Bartholomew made several suggestions of potential sites that would not "offend the white community." Still, the white community defeated the firm's most attractive recommended site, and the Bartholomew proposal for a 3,000-acre "river bottom" site was rejected by the African American community. The site finally agreed upon lay in an unincorporated area ten miles north of downtown Dallas.[9]

To contextualize Hamilton Park better it is also necessary to address the perceived housing crisis within the African American community. This was precipitated by three issues: the City of Dallas's demolition of African American housing for the expansion of Dallas Love Field; the 1950s terrorist bombings of African American–owned housing as they moved into traditional white neighborhoods in South Dallas; and growth in the African American community, which increased by more than 30,000 people between 1940 and 1950. Ultimately, Hamilton Park was far enough away to not cause discomfort to the white community but close enough to enable African American workers to meet employers' needs. At the urging of A. Maceo Smith of the FHA and John W. Rice of the Black Chamber of Commerce, Lynn Landrum, a columnist with the *Dallas Morning News*, wrote an article calling for a housing site where African Americans could "run the development as they wanted to."[10]

With the Hoblitzelle Foundation loan and private financing, the Dallas Chamber of Commerce and the Dallas Citizens Council established an association in 1951 to build Hamilton Park. The Hamilton Park subdivision was dedicated in October 1953 and formally opened in May 1954— the same month and year of the historic Supreme Court decision *Brown v. Board of Education*. The summer of 1954 also saw Dallas hosting the national meeting of the NAACP, with more than seven thousand delegates. In addition, St. Paul Hospital announced the appointment of the first African American doctors to its staff.[11]

The Hamilton Park subdivision was virtually completed in 1961 with 742 single-family homes, an apartment complex, a shopping center, public park, and a local public school. A one-bedroom house sold for $8,500. A four-bedroom house sold for approximately $11,000. Showing the pride of the African American residents of Hamilton Park, they immediately moved to organize themselves to preserve the value of their unique community. The three largest churches in the area—Hamilton Park United Methodist Church, the First Baptist Church of Hamilton Park, and the Hamilton Park Church of Christ—addressed the spiritual needs of the community and helped its members create several nonprofit organizations to enhance civic life. The first of these was the Civic League, a community-based organization devoted to the preservation and improvement of the physical needs of the community. The Civic League also acted as an intermediary with government agencies for the betterment of Hamilton Park. One of its most important community projects was the development of Willowdell Park, a 21-acre site. Starting in 1957 the Civic League also petitioned the City of Dallas park and recreation board to install playground equipment, a wading pool, a lighted baseball diamond, and a fence. Over time, the City of Dallas planted trees, built sidewalks inside the park, and installed a lighted tennis court.[12]

Another significant activity and one that involved the entire community of Hamilton Park was organization of community-wide events that focused on property maintenance, controlling stray dogs, and a competition through a lawn beautification program.[13] In addition, also in the 1950s, the Hamilton Park Interorganizational Council (IOC) formed, which became an umbrella organization in the community that attempted to coordinate all the civic, charitable, religious, educational, and cultural groups. It found its success in two primary areas: voter registration and education; and the desegregation of the Hamilton Park School, a facility of the Richardson Independent School District. Impressively, the IOC further helped to establish the renowned Pacesetter program in 1975 whereby the parents of non–African American children volunteered to send their children to Hamilton Park rather than other schools in the Richardson district. Also, by promoting candidate forums and endorsements, the community was able to become more active in local politics. Former governor John Connally, for example, came to a forum in his bid for the governorship in 1968. Once candidates were endorsed, the IOC gave broad distribution of its candidates list and encouraged others to support its slate. In fact, Hamilton Park had one of the highest voter percentage turnouts in Dallas County (during this time, poll taxes were collected at the churches, until they were declared unconstitutional).[14]

After 1968 the Hamilton Park community lost some of its attractiveness because of larger social and economic gains in the African American community throughout Dallas and the nation. Though nowhere near the so-called cure-all they portended to be, these general improvements are attributed to the passage of the Civil Rights Acts of 1964, 1965, and 1968. These three pieces of legislation provided for open public accommodations, elimination of certain voting restrictions, and open housing to all, regardless of race, religion, and national origin. African Americans thus began to move from South Dallas to Oak Cliff in southern Dallas. Several original Hamilton Park residents moved to larger and more expensive housing in the McShann Road housing development and to Richardson and Plano.

In 2014, Hamilton Park turned sixty years old. As with all communities, time brought about many physical changes. The Hamilton Park First Baptist Church, for example, had moved to Richardson. The community was beginning to show signs of deterioration as well. Many of the pioneer residents no longer lived in Hamilton Park, and reports of violence, property crimes, and thefts were increasing. Over the previous twenty years, the community had experienced a decline in voting as well—a sign of losing community interest and participation.

Though Hamilton Park is the most famed of the African American suburban areas in Dallas, precisely because it was better off than other African American neighborhoods, off McShann Road in North Dallas was another story of development in the 1950s and 1960s by and for middle- to upper-class African Americans. Parallel to the white civic community initiatives to provide decent, affordable, and sanitary housing for African Americans in Dallas, a group of professional African Americans took it upon themselves to develop suitable housing in an unincorporated area of what is now known as North Dallas. The project would become known as the McShann Road housing development, an area just south of the LBJ Freeway bound on the west by Montfort Road and on the east by Preston Road. The overall area is known as Preston Hollow, one of the wealthiest communities in Dallas.[15]

The McShann Road housing development was built on land owned by freedmen dating back to the Civil War. The owners were Will and Diza McShann. Their land was a part of hundreds of acres owned and farmed by African Americans. This area is now known for the Valley View Shopping Center and its environs. The historic African American Mt. Pisgah Church building still stands at Preston Road and Valley View Road. According to Jessie Arnold, an eighty-five-year-old historian and resident of the area, he worked the land and found the money to build his home as part of the subdivision. He acquired his land in 1967 and moved into his home in 1971. Robert Prince, a physician and local historian, remembers that the area was planned as an upscale and affluent African American neighborhood when Dallas did not have such neighborhoods. Planned as a community to attract doctors, over time the area attracted attorneys, teachers, engineers, entrepreneurs, and executives. The area now houses residents who are employed in or retired from various professions.[16]

Barbara Watkins, a longtime resident and former director of the Parkland Hospital Foundation, remembers when everyone in the neighborhood knew each other and conversed as they left their houses for work. She also remembers how impressive it was on weekends to see the owners working on their well-maintained lawns and improving their property. Chris Flowers, who lived all of his young life on McShann Road, remembers when young couples would visit the homes on social occasions and especially for debutante parties.[17]

Since 1995, the McShann housing development has been moving in a different direction. The first orthodox Jewish family moved into the neighborhood around 1994. Since then, three Jewish synagogues have been built near the neighborhood and the Jewish population has expanded. The McShann community has become a unique, racially mixed neighborhood with stately homes, shaded trees, and manicured lawns.

The more than twenty-five suburban communities north of Dallas first attracted new residents for growth because of "white flight" from Dallas in the 1970s and 1980s, mainly to avoid public school integration.[18] This wave of suburbanites north of Dallas followed several corporate relocations to Legacy Park and other designated business and industrial parks nearby. These suburban communities have thus accounted for most of the job growth there. Still, suburban development in the area dates back to the 1950s as well. Its roots began when Texas Instruments closed its semiconductor facilities in Dallas and moved to a 300-acre site along Central Expressway in Richardson. In the decade of the 1950s, Richardson grew by 120 percent. Since then, as Paul J. P. Sandul notes in his chapter, the Plano and Frisco areas have been two of the fastest-growing areas in the United States.

The decade of the 1960s saw the Southwest Center for Advanced Studies become the University of Texas at Dallas in Richardson. It famously led to the startup of several electronic and industrial companies by alumni of Texas Instruments and Collins Radio, and MCI Telecommunications moved to form the Telecom Corridor in Richardson. Not to be outdone, the relocation of EDS Electronics from Dallas to Plano in the 1980s gave birth to Legacy Business Park and attracted other industrial relocations such as Frito-Lay and J. C. Penney. The opening of the Hall Office Business Park north of Legacy Park in the 1990s led several national firms to build offices in Plano, including the Dr. Pepper Group and Alcatel-Lucent. At the turn of the twenty-first century, suburban shopping boomed, including The Shops at Willow Bend in Plano and Stonebriar Center in Frisco.

African Americans now living in the suburbs north of Dallas have a variety of origins: some were pioneer families who had lived in the rural communities before they became cities (this is particularly true of Plano); African American executives came along with industrial relocations such as J. C. Penney and Frito-Lay; professional athletes came to join such sports teams as the Dallas Cowboys, Dallas Mavericks, and Texas Rangers (for example, the largest house in Prosper, Texas, is 29,000 square feet and owned by a former Dallas Cowboy); and several nationally known entertainers have made their homes in the northern suburbs. In addition, since 1968 a small percentage of African American professionals have moved from Dallas to these suburban communities.

From the 1950s through 2010, the expansion of Central Expressway, the construction of the North Dallas Tollway to Frisco, and the building and expansion of the Dallas Area Rapid Transit (DART) system aided growth. The larger suburban communities north of Dallas currently market themselves in the familiar suburban ethos as residential communities with small retail shopping, professional services, and medical facilities. In the summer of 2014 the Dallas Cowboys organization also broke ground in the Frisco area to build a new headquarters, practice field, and world-class office with retail spaces as well. A new performing arts center has recently been built in Richardson, and Plano has had several new school buildings built and became a statewide rival in high school sports. The Allen Independent School District recently built a $60 million high school football stadium with a capacity of more than 18,000. Plano elected its first African American mayor in 2014.

Suburban communities east of Dallas such as Mesquite focus on retail growth because of their common boundary with Dallas. Because of this success and the improvements in its tax base, Mesquite has developed a professional rodeo stadium and conference center, an arts center, and a state-of-the-art police station. It annually shows its pride with the Mesquite Real Texas Festival as well, featuring a BBQ cookoff, classic car show, carnival, and, of course, a rodeo. More noteworthy here, however, is that Mesquite's African American population grew from 1 percent in the 1970s to more than 10 percent in 2010. Sunnyvale, a small community along Interstate 20 and Beltline Road, is also a high-income community with large houses and lots with well-kept lawns. Yet, because of Sunnyvale's expensive building requirements, the community has been cited for maintaining discriminatory policies. For instance, a federal court order recently forced the city to provide apartments for mixed-income groups. Rockwall meanwhile has developed as a suburban city on the lake because of its proximity to Lake Ray Hubbard, which is owned by the City of Dallas and is Dallas's primary water supply. Rockwall focuses on residential development, restaurants, and water sports activities.

Dallas suburban communities to the south of Dallas are Lancaster, DeSoto, Duncanville, and Cedar Hill. These suburban areas were developed as so-called bedroom communities and, for all practical purposes, remain so with very little growth in retail, commerce, or industry. The employment base for residents of these communities is in Dallas and the northern suburban communities. The Dallas–Fort Worth airport and parts of Arlington, Irving, and Los Colinas also provide employment for many of the African American residents of these communities. Because African Americans are heavily represented in government employment, many of the residents in the southern suburbs work for the Dallas Independent School District, the federal government, the City of Dallas, Dallas County government, Baylor Hospital, and Parkland Hospital.

The overall growth in the African American middle class in the Dallas area suburbs has come as a result of their finding professional employment in education, government, and the medical fields. The engine for growth in the southern suburbs is different from that of the northern suburbs, however. Specifically, the problem with income disparities is one of the causes. This is compounded by a lack of a strong economic base in the southern suburbs, and programs that promote real estate on television rarely market the southern suburbs (an example is the "Hot on Homes" television program).

Despite real differences from one Dallas suburb to the next, a typical example of local African American residents moving to suburbia is the case of the Lionel Woods and his family.[19] Woods, now over eighty years old, was born and raised in Winnsboro, Texas, in Smith County. He was raised in the Sand Flat community and graduated from a local, segregated, Rosenwald School for African Americans. His parents were sharecroppers with a one-fourth share. As school board trustee for the Jackson Colored School his father was actively involved in the community. Woods remembers being an active participant in the 4-H Club at Jackson Colored School and winning prizes and selling eggs and pigs. For many years his mother was the president of the Colored Branch of the 4-H Clubs in Smith County and took an active role in the Smith County fair. Woods finished ten years of schooling in 1946 and over the next sixteen years earned a college degree and enlisted twice in the U.S. Army.

Leola Woods was born and raised in Quitman, Texas, in the Muddy Creek community located in Wood County. She graduated from W. B. Carter Colored School and immediately enrolled in Jarvis Christian College in 1943. Woods graduated with a major in history and a minor in English. Her first employment as a teacher was in the West Texas town of Spur. She eventually got her master's degree from Prairie View A&M College. The Woods were married in Quitman, Texas, in 1957 and built their first home in Winnsboro in 1959.

The Woods moved to Dallas in 1963 for Lionel to take a job with Texas Instruments. Leola became an elementary teacher at William Brown Miller in Oak Cliff (southwest Dallas). She later taught at several other schools and retired in 1974. When the Woods first moved to Dallas, they lived in an apartment on Southerland Street in South Dallas. They then purchased their first home in Dallas in 1965. Later, they moved to the Singing Hills community and then to Franklin Street in west Oak Cliff, where they stayed for nearly thirty years. The Oak Cliff area is where most middle-class African Americans live within the city of Dallas.

The Woods moved to the suburb of DeSoto in 2003 after purchasing a lot and having a custom home built. They now live in a one and one-half story, 4,000-square-foot home in a gated community at Thorntree Country Club. They enjoy a house with five bedrooms and five baths that has a market value in excess of $500,000. The Woods have enjoyed a remarkable career and lifestyle. They adopted two daughters from a deceased relative and are now raising two adopted grandsons. In the course of their careers they were able to build two three-bedroom houses in East Texas for each of their mothers. They also helped to organize Sunbelt National Bank and made substantial investments in other projects. Their biggest disappointment was losing over $600,000 in the collapse of Sunbelt National Bank, an African American–owned and –operated financial institution.

Lionel Woods achieved his success in an area where few African Americans excel—as an entrepreneur. He organized and owned three Shell service stations in South Dallas and Oak Cliff. Before retiring he led the local Texas Shell district in volume of gas sold and trained many other African Americans in service station management. Perhaps clearly the Woods story represents many African American families that started in East Texas during the Jim Crow era and moved to Dallas to realize their dreams, find employment, and later moved on to suburbia.

Although upper middle-class African Americans willingly move to suburbia, they often maintain their Dallas-based church affiliations and their social networks with fraternities, sororities, and other social and civic organizations.[20] An informal survey of upwardly mobile African Americans revealed several reasons for moving to suburbia. Fred Myers, owner of a popular barbershop and other business entities in Dallas, said that his main motive for moving to DeSoto more than ten years earlier was to "get more house for the dollar." Letitia Hughes, a real estate agent in Dallas, moved to DeSoto because of its schools for her young family. Annie Jones Evans, a graduate of Dallas Independent School District, bought a house in Arlington after she graduated from the University of Texas at Arlington and found the area to be a satisfying place to live. Her vocation is the management of a nonprofit organization in Dallas. Mark Proctor, a mid-level manager for the City of Dallas, saw the city of Cedar Hill as a wholesome place to raise a growing family. Frankie Washington, a successful funeral home owner and director in Dallas, is among a few young African Americans who have moved from Dallas and beyond the inner suburban areas to build a new 5,000-square-foot house, helping to create an outer suburban area south of Dallas.[21]

Although the southern suburban areas had their origins in the 1800s, they did not take on their more suburban bedroom status until the 1950s. The four principal southern suburbs are DeSoto, Cedar Hill, Lancaster, and Duncanville. Among these southern suburbs, DeSoto appears to be more in line with the posher suburbs in the north. For example, in the Frost Farms community lots sold for more than $50,000 in the 1990s. A successful country club also anchors the community, and some houses sell for over $1,000,000. Over the past decade southern suburbs have even received several honors and awards. In 2006, for example, the National Civic League named DeSoto an "All American City." In 2005 and 2006, Lancaster was a finalist for the "All American City" award, and in 2007 the National Arbor Day Foundation designated Lancaster a "Tree City USA." The Texas Historical Commission designated the city of Duncanville an "Official Main Street City," and, earlier in its history, Cedar Hill served as the temporary county seat of Dallas County.

Unlike the northern suburban communities that provide housing for hundreds of professional athletes, the southern suburban communities can lay claim to only two well-known and retired professional athletes— Mean Joe Green of the Pittsburgh Steelers and Tim Brown of the Oakland Raiders. They both were educated and lived in Dallas before their football careers. Still worth noting, however, according to the 2010 census, demographically, African Americans accounted for 68 percent of the total population in Lancaster and DeSoto and 29 percent and 51 percent of the total populations in Duncanville and Cedar Hill, respectively. Duncanville was the only southern suburban community with a sizable Hispanic population, at 35 percent (in the 1970s, Duncanville was actually known as a "white flight" area).

Because of these sizable African American populations in the southern suburbs, several African Americans hold elected positions on the boards of education and the city councils. In fact, African Americans serve as mayor, city manager, and chief of police in Lancaster. In the area of public education the Duncanville Independent School District outperforms all surrounding public school districts, including Dallas (thirteen of seventeen schools are rated exemplary or recognized by the Texas Education Agency).[22]

The suburban migration out of the city of Dallas—of all ethnic and racial groups—has had positive and negative consequences. The changes in demographics over the past forty years have caused a positive change in Dallas politics and the manner in which the city is governed. Yet the tax base has been negatively affected as wealthier citizens take both their taxes and spending dollars to the suburbs. Although the population of Dallas is static at around 1.2 million, the concentration of poverty is growing and African American neighborhoods have suffered from a lack of both leadership and capital flow. The population that Dallas lost over the past twenty years has been replaced with an equal number of migrants who are Hispanic and with less income. This translates into the increasing local tax burden being carried by fewer and fewer residents of Dallas. If this trend is allowed to continue, at some point Dallas will become less competitive with its suburban communities for industrial relocations and conventions.

The Mayor's Taskforce on Poverty in July 2014 reported that, in year 2000, 10 percent of Dallas's population was considered poor and 4 percent of the total population lived in neighborhoods of concentrated poverty. The taskforce further reported that in the year 2012 more than 21 percent of Dallas was considered to be poor by federal guideline standards and 10 percent of Dallas's poor population lived in neighborhoods of concentrated poverty; as well, the number of city council districts with concentrated poverty neighborhoods had increased from seven in 2000 to nine in 2010. This clearly shows the spread of poverty and so-called blight. Poverty in Dallas is disproportionately concentrated in minority groups who represent over 50 percent of the general population.

The Mayor's Taskforce made six recommendations. The three most significant recommendations appear to be, first, that the city leadership recognize the current problem and project future trends. The taskforce recommended that the city establish within local government an Office of Community Opportunity to serve as a super-agency to coordinate activities focused on helping the poor. Second, the taskforce called for the establishment of a $1 million investment fund to finance the creation and expansion of tax aid financial centers to give the working poor "know-how" to take advantage of the federal income tax credit program. It is estimated that 40,000 low-income Dallas families qualify but are not taking advantage of the program. This translates into over $55 million annually in lost income to the local working class. Finally, the taskforce called for the city to require a higher minimum wage for the working poor to be able to have a decent lifestyle.[23]

All of the aforementioned solutions are focused on what the City of Dallas can do to turn around the increasing poverty trends. Realistically, to develop a "level playing field" between Dallas and its suburban communities, the public and private leadership will need to consider proposals for "income sharing" between Dallas, the central city, and the suburban communities in Dallas County and beyond. Successful models for this approach can be found in Minneapolis–St. Paul and Nashville–Davidson County, Tennessee. If this is not done, Dallas at some point might be faced with instituting defensive measures that will cause conflict with the suburban communities, especially in North Dallas. Among the defensive measures that Dallas might consider is a toll tax for commuters to enter the city; residential requirements for all city workers; an increase in the city water distribution charges for all areas outside of Dallas; and a surtax on all visitors to Dallas art, cultural, and recreational facilities. Whatever the case, suburbia in Dallas, including African American suburbanization, is both booming and strengthening.

NOTES

1. *Dallas Morning News*, map showing the central city and its physical relationships to its suburban communities, February 9, 2014, 11C.

2. Andrew Wiese, *Places of Their Own: African American Suburbanization in the Twentieth Century* (Chicago: University of Chicago Press, 2004).

3. Theodore M. Lawe, "Racial Politics in Dallas in the Twentieth Century," *East Texas Historical Journal* 46, no. 2 (2008): 30; Stephen Grant Myer, *As Long as They Don't Move Next Door: Segregation and Racial Conflict in American Neighborhoods* (New York: Rowman and Littlefield, 2000), 20, 28.

4. Lawe, "Racial Politics," 30–32.

5. Dallas Black Chamber of Commerce, *Changing the Profile of Dallas* (Dallas: Dallas Black Chamber of Commerce, 2005), 6–12; *The Handbook of Texas Online*, s.v. "Hall of Negro Life," www.tsha.utexas.edu/handbook JonlineJarticle~HH/pkhl.html (accessed September 12, 2005).

6. Jim Schutze, *The Accommodation: The Politics of Race in an American City* (Secaucus, NJ: Citadel Press, 1986), 77–78; Donald Payton, "A Concise History of Black Dallas since 1842," *D Magazine* 25, no. 6 (June 1998): 26–31.

7. Lawe, "Racial Politics," 35–36.

8. William Wilson, *Hamilton Park: A Planned Black Community in Dallas* (Baltimore: Johns Hopkins University Press, 1998).

9. William Wilson, "The Negro Housing Matter: A Search for a Viable African American Residential Subdivision in Dallas, 1945–1950," *Legacies: A History Journal for Dallas and North Central Texas* 6, no. 2 (Fall 1994): 28–311; Wiese, *Places of Their Own*, 196–208; Schutze, *Accommodation*, 109.

10. Wilson, *Hamilton Park*, 25–31, 62–63.

11. Wilson, *Hamilton Park*, 33–54; Lawe, "Racial Politics," 33; Robert Prince, *A History of Dallas from a Different Perspective* (Austin: Nortex Press, 1993), 111.

12. Wilson, *Hamilton Park*, 67.

13. Wilson, *Hamilton Park*, 93–104.

14. Wilson, *Hamilton Park*, 54–147.

15. The following review of the McShann Road development is from Kris Scott, "Turning a Corner," *Advocate Magazine*, January 1, 2004, https://prestonhollow .advocatemag.com/2004/01/01/turning-a-corner (accessed September 20, 2014), along with authors' interviews, as noted.

16. Interview with Robert Prince, September 25, 2014.

17. Interview with Barbara Watkins, September 14, 2014; interview with Chris Flowers, August 30, 2014.

18. Tracy Curtis, "Is White Flight Ruining the Dallas Schools?," *D Magazine*, August 1977, www.dmagazine.com/publications/d-magazine/1977/august/is-white -flight-ruining-the-dallas-schools.

19. Interview with Lionel and Leola Woods in their home in DeSoto, TX, January 11, 2013.

20. We make this assertion after discussions with members of fraternal organizations and social clubs over a two-year period.

21. Interview with Fred Myers, July 9, 2014; interview with Letitia Hughes, August 27, 2014; interview with Annie Jones Evans, July 9, 2014; interview with Mark Proctor,

July 16, 2014; interview with Frankie Washington, January 13, 2013.

22. For information concerning Duncanville, Desoto, Lancaster, and Cedar Hill, see the following websites: www.duncanville.com; www.citydata.com/city/desoto -texas.html; www.lancaster-tx.com; and www.cedarhilltx.com.

23. "Taking Aim at Poverty," *Dallas Morning News*, August 20, 2014, 1, 5.

{ 5 }

THE CZARS OF CONCRETE

THE TEXAS HIGHWAY DEPARTMENT'S ENGINEER-MANAGER PROGRAM AND THE DEVELOPMENT OF URBAN EXPRESSWAYS

TOM McKINNEY

One of the key factors in suburbanization is the development of dependable transportation facilities. Transportation facilities, whether they be streetcars, automobiles, or a commuter railroad line, make suburbs possible by opening up undeveloped land for suburban development. In short, suburban real estate developers seek to develop land that can be accessed by people living in the central city so that the prospective residents can commute to and from the central city.[1] The manner in which suburban development occurs is dependent on the type of transportation facility or facilities that provide access to it. For example, Austin's Hyde Park has a wholly different layout than Houston's River Oaks; Hyde Park was constructed around streetcar access in the 1890s, whereas River Oaks was designed for automobile access by the 1920s.

Largely speaking, the vast majority of Texas suburbs were constructed during the post–World War II era, a time when the personal automobile was undeniably the dominant mode of transportation. The actual trends toward suburbanization and decentralization originated in the Lone Star State in the 1920s, but both of these came into full bloom after World War II. The postwar era surely encouraged the development of automobile suburbs in both Texas and the nation, for the price of gasoline was low and the federal government embarked on the construction of the nation's interstate highway system, subsidizing the construction of both urban freeways and "Little Boxes on the Hillside."

The major metropolitan areas of Texas decentralized during the 1920s as the automobile quickly became the primary form of transportation. For example, drive-in stores, or "rolling patronage," as it was called at the time, sprang up all over Houston in the 1920s to cater to automobile commuters, many of whom drove to and from the downtown area as many as five times a day.[2] The development of cultural attractions outside of Houston's core, such as the Rice Institute [now University], the Museum of Fine Arts, and Hermann Park, marked the origin of Houston's present multinuclear form as well as its automotive dependence, for all of these places are located well outside the urban core of the city. This is supported by the fact that 90 percent of Houston's development occurred during the era of the automobile.[3]

Decentralization also occurred in Fort Worth during the 1920s. The Pig Stand, the first drive-in restaurant chain in the United States, began with a franchise on the highway between Fort Worth and Dallas. Montgomery Ward also opened the first business outside of the Fort Worth core in 1928, heralding the beginning of the construction of shopping centers outside the city's downtown. The citizens of Fort Worth also witnessed the appearance of traffic policemen, speed limits, and downtown curbside parking. The automotive amenities led the people of Fort Worth to "embrace freeway construction the same way they had railroad building in the previous century."[4]

San Antonio's decentralization also began in the 1920s, with the automobile allowing the construction of "high-end suburbs" on the edge of this former walking city. The suburban communities of Woodlawn, Terrell Hills, and Olmos Park were all constructed by the wealthier members of the San Antonio community, resulting in a diminishing supply of tax dollars into the city's coffers. The development of these suburbs also renewed the Alamo City's commitment to the automobile as these politically powerful elite suburbanites encouraged the further development of roads, freeways, and other pieces of urban infrastructure. These improvements in turn encouraged additional suburban growth, more reliance on the automobile, and an increasing number of suburbanites who enjoyed all San Antonio had to offer without contributing taxes for its maintenance.[5]

Dallas was no exception to 1920s decentralization. Dallas elites incorporated Highland Park in 1913 to the north of the central city, and other developments followed in its wake. Dallas real estate developers also began locating shopping centers along the city's main arterial routes to help serve suburbanites. In fact, Highland Park Shopping Village, the first of these decentralized shopping centers (opened in 1931), was added to the National Register of Historic Places in 1986 because of its significance in the evolution of the modern shopping center.[6]

The decentralization of Texas's major metropolitan areas generally occurred hand in hand with an aggressive annexation campaign of suburban communities (see Andrew C. Baker, Robert B. Fairbanks, and Andrew Busch in this volume for more about annexation and incorporation in Texas). Houston, the undisputed champion of annexation in the state, used its freeway routes as pathways of annexation.[7] The rapid suburban growth that occurred along the fringes of this city during the postwar era contributed heavily to the escalating number of automobiles that poured onto Houston's streets and emerging freeway network as federal monies created the core of the city's freeways. Urban congestion was also exacerbated by the city's aggressive postwar annexation policy, which doubled the size of the city in 1949 and made Houston the largest city in the South. By annexing such a large area, however, Houston created a dire need for transportation facilities, and the sheer size of the metropolitan area presented highway planners with a daunting obstacle. Mayor Oscar F. Holcombe resurrected the Office of City Planning in 1937, and by 1943 this office was responsible for the city's major thoroughfare plans.[8]

The Bayou City had remained steadfast to pro-growth politics since the Allen brothers founded it, but Mayor Holcombe's administration took this policy to previously unconsidered heights, using the power of annexation effectively to conquer the city's fear of becoming "hemmed in" by encircling incorporated subdivisions. This "single-minded drive" to annex as much of the region as possible was considered by the overwhelming majority to be good policy. Holcombe felt that this was a way for Houston to assume a leadership position among America's great cities, and few Houstonians doubted him.[9]

Though the possibility of having the city encircled by numerous incorporated suburbs was effectively dismissed by the titanic 1949 annexation, the city also burdened itself with an ever-increasing demand for city services in the newly annexed areas. The city also assumed the debt incurred by the annexed area, adding a further burden to its coffers.[10] In short, Houston's annexation policy allowed the city to enjoy unfettered growth, but at a huge price tag; providing city services and transportation facilities and assuming the debt of marginally populated areas strained the city's pocketbooks.

Houston's aggressive annexation policy coupled with rapid postwar suburbanization resulted in increasing congestion on formerly rural and urban streets.[11] Railroad traffic also contributed to the problem as trains blocked major thoroughfares for longer and longer periods of time.[12] Mayor Holcombe's planning department included a division of traffic and transportation, showing his awareness of the need for adequate traffic management.[13] Under his leadership, the City of Houston introduced traffic control measures, gaining the distinction of becoming the first city in the United States to use sequential traffic lights and parking meters.[14]

Annexation created a grim financial picture for Houston, however, especially when one considers the overwhelming total cost of the extension of water and sewage facilities, the cost of police and fire protection, the creation of educational facilities, the assumption of new debts, as well as the cost of new transportation facilities. Previously Houston's arterial and residential streets were paved by funds from the city and abutting property owners. The property owners paid 90 percent of the cost of permanent paving and the city paid 10 percent (100 percent at intersections) and for all storm sewers. This was not the case with the Gulf Freeway, however, which was financed by city, state, and federal entities. This arrangement did not apply to many of the projects in the newly annexed areas. The City of Houston was not able to keep up with the ever-increasing demand for new traffic facilities, as its rapid growth outstripped its ability to finance and construct new transportation facilities. As one reporter said, "Houston will have a problem in providing enough through streets to carry traffic into all sections of the city."[15]

Fort Worth also embarked on an aggressive annexation campaign throughout its history, so much so that the city's growth has been outward toward its periphery since the city was first founded in 1873. The 1956 study that revealed the outward nature of Fort Worth's growth concluded that this led many outlying suburban communities to incorporate to protect themselves from annexation. The author of the study, Robert H. Talbert, commented that "this often creates problems of duplication of services, conflicts of legal authority and raises the issue of whether the residents of the independent suburb are paying for services they use in the central city."[16]

As Robert B. Fairbanks and Philip G. Pope discuss in their contributions to this volume, Dallas also embarked on an aggressive annexation campaign during the postwar era, but this ended when the city became completely surrounded by incorporated suburbs.[17] This is the scenario that formed the stuff of Houston mayor Oscar Holcomb's worst nightmares, and Houston's massive 1949 annexation solved this problem for Houston.

For its part, Dallas shifted its pro-growth focus from outward growth to upward growth and, in 1999, it was the fifth-fastest-growing city in the nation with its new citizens taking up residence in apartment complexes and high rises.[18]

The passage of the Federal-Aid Highway Act of 1944 forever changed the way roads are planned and constructed in the United States. The act dramatically increased the amount of federal money budgeted for highway construction, an amount that continually increased during the postwar era. More important, the act marked the first time that the federal government provided funds for the construction of urban roads, with previous highway appropriations having been rurally focused.

The new law defined an urban area as "an area including and adjacent to a municipality or other urban place, or five thousand or more, the population of such included municipality or other urban place to be determined by the latest available Federal census. The boundaries of urban areas, as defined herein, will be fixed by the State highway department of each State subject to the approval of the Public Roads Administration."[19] This stipulation was specific to urban projects; the funding formulas for primary and secondary highways took into account state land area, farm-to-market road mileage, and other variables. The government examined population only in determining an urban center's eligibility for federal highway dollars.

Previously, the federal government had considered urban roads to be the exclusive domain of the municipalities they served, and, as a result, previous highway aid acts focused exclusively on rural road construction and maintenance. The 1944 act, however, extended state highway departments' jurisdiction into cities. By recognizing the need for urban highways, the federal government also realized the financial consequences of urban congestion.[20] By earmarking funds for the construction of urban highways, the federal government sought to eliminate congestion in urban centers by ensuring that financing of such projects was available to all cities. This also served an economic function. The American trucking industry experienced a boom during the postwar era, greatly aided by conservative exemptions in antitrust laws to boot (e.g., Reed-Bullwinkle Act, 1948).

Thomas H. MacDonald, the head of the United States Bureau of Public Roads, realized that this extension of state and federal power into a previously local domain could not be carried out in a heavy-handed manner. To ensure that this new program was not bogged down because of politicians' jurisdictional squabbling, MacDonald believed that the selection of

highway routes in urban areas "should represent the best composite thinking of the city, State and Public Roads Administration personnel. Certainly
cooperation and consultations with cities are desired. The approach should
be conservative and it should be emphasized that our first step is to identify
desirable routes and prepare maps showing tentative selections."[21] This was
particularly key to road construction in the urban centers of Texas, where
state law maintained that it was the city's responsibility to obtain the right-
of-way for a roadway and surrender it to the state highway department.
This ensured not only that the highway department was unable to force
a roadway through a city but also that it fully understood the needs and
concerns of the urban population the department was attempting to serve.
The construction of urban roads in Texas, therefore, required the building
of a coalition of pro-roadway officials from all levels of government.

Congress passed the Federal-Aid Highway Act of 1944 in "anticipation
of the transition to a postwar economy and to prepare for the expected
growth in traffic."[22] MacDonald felt that Congress passed the measure
to shore up employment in an effort to keep the economy stable during
the economic conversion from war production to peacetime production.
Writing to Texas congressman Wright Pateman in 1949, he argued that
this reasoning was wrong. Highway construction, as it turned out, was
not needed to drive the American economy while the industrial sector
regained its footing. Furthermore, he found that repopulating state highway departments was difficult after the war, because of the inability of the
departments to compete with private companies.[23] These circumstances
plagued many state highway construction projects during the postwar
era as returning veterans sought higher-paying jobs in the private sector.
More important, shortages of necessary materials delayed the completion
of many postwar highway projects.

Aside from providing funds for urban projects, the Federal-Aid Highway Act specifically called for the states to designate a coherent system
of highways. This was a revolution in American road planning. Rather
than designate a separate system for state highways, rural roads, and city
streets, the 1944 act forced states to look at their transportation needs from
the broadest scope possible and create a balanced system of roads. By tying
long-range planning to road development, the new legislation forced state
highway departments to examine their state's transportation needs seri-
ously. This stipulation is implicit in federal legislation since World War II
and is the most important idea in modern highway planning.[24]

The Federal-Aid Highway Act of 1944 provided for a 40,000-mile "National System of Interstate Highways" to connect "principal metropolitan areas, cities, and industrial centers to serve the national defense, and to connect suitable border points with routes of continental importance in the Dominion of Canada and the Republic of Mexico." This system of interstates was subject to the approval of the Bureau of Public Roads, which allowed the federal government to propagate coherence and enforce design standards. The Federal-Aid Highway Act of 1956 amended the 1944 act by increasing its funding provisions. This act launched the construction of the U.S. system of interstate and defense highways, the largest public works program undertaken in this nation. The 1944 act, though endorsing the interstate system, did not provide the necessary money needed for construction.[25]

The Texas road system was a causality of World War II. The Texas highway department, headed by Dewitt C. Greer, "virtually suspended" highway development in Texas during the war. What was constructed was severely limited because of the need to conserve war materials and was generally focused on defense needs; the department's low point was 1944, the year that the department began work on only three hundred miles of roadway.[26] Besides total neglect by the highway department, Texas highways also suffered because of heavy use by both civilian and increasing military traffic. Beginning in 1940, the U.S. Army conducted a series of war games known as the "Louisiana Maneuvers" across the Texas-Louisiana border. The army discovered that "only one-fourth of the East Texas roads used met federal military standards, and hundreds of miles of roads and many bridges suffered damage from the heavy traffic."[27] The substandard road system of rural eastern Texas, as well as many other highways throughout the state, fell into disrepair because of neglect and military use.

Division engineer J. A. Elliott of the Bureau of Public Roads found that this was common in the United States. In 1949 he commented:

Prior to World War II the increase in the number of motor vehicles in service coupled with the increased use of the motor vehicle caused a public demand for a large mileage of highways usable immediately. The only way this mileage could be obtained was by sacrificing desirable standards. During the war, highway construction aside from that considered essential to the war effort was practically nil. At the end of the war the States found themselves with an increased mileage requiring reconstruction and higher maintenance on all mileage.[28]

Despite the fact that the Texas highway department did not initiate many projects during the war, it did not spend the war idly. Suffering from a severe lack of manpower and materials, the department used the break in construction to focus on planning initiatives in hopes of a postwar renaissance. Beginning in 1944, it conducted a survey to study transportation problems in Houston, Dallas, Fort Worth, and San Antonio and enlisted the aid of "nationally-known highway planners." These planners concluded that the urban environment of these cities would not allow the sufficient widening of streets to relieve the mounting traffic congestion caused by increased suburban development and automobile ownership. As one engineer pointed out, "A 200-foot street, as required for an expressway, ruins practically the whole block."[29] The planning survey discovered "that nearly one-half the highways in the state were ten or more years old, while in 1940 only one-third had been. Aging and deterioration of structures and surfaces were increasing rapidly, since little had been done to renovate the system during war years."[30] The federal government openly endorsed this policy of neglect, for it wanted nothing to interfere with the war effort. To ensure the compliance of all of the state highway departments, it drastically cut all aid for nonmilitary highway projects. This virtually halted highway construction in the United States during the war.[31]

While overseeing the drafting of postwar highway plans for Texas, Greer also did all he could to ensure that funds were available for construction once the war ended. He invested much of the highway department's budget in treasury bonds, "for postwar reconstruction of the state's highways." By the end of the war, Greer's investment gave the department a $30 million dollar nest egg that was drawing 0.875 percent interest and was available immediately for highway repair and construction.[32] This fact, coupled with a coherent statewide plan and renewed federal funding, allowed Texas to proceed aggressively with its postwar highway program.

Greer created a series of special posts called "engineer-managers" to oversee the construction of urban highways in Texas. The posts existed in the four major cities of the state at the time (Houston, Dallas, Fort Worth, and San Antonio), and the engineer-managers were given the sole responsibility for planning and overseeing the construction of urban expressways. The highway department empowered this new position to make rapid decisions concerning urban expressways, showing that the highway department realized the importance of these routes and desired to finish them rapidly. William James Van London was appointed on June 1, 1945, by the department to supervise the preparation of plans, specifications, estimates, and the construction of Houston urban expressways.[33] The Gulf Freeway, the first toll-free highway constructed after World War II in the United States, was his first project.

The engineer-manager program was highly successful and received praise from the Bureau of Public Roads. D. W. Loutzenheiser, a bureau official, commented:

> The Engineer-Manager plan being used in Texas obviously is a successful method. The key to this success seems to be the delegation of authority to the man in charge which permits him to make decisions and act rapidly to get the projects underway. It appears that the city officials are well pleased with the progress being made and the type of facilities provided. . . . Inasmuch as the cities are furnishing this will determine the scheduling of next sections.[34]

Division engineer J. A. Elliott of the Bureau of Public Roads said in 1949, "Progress is being made in other States but not on such a large scale as in Texas."[35]

Dallas, Houston, San Antonio, and Fort Worth all sought to become automobile cities, and each did its best to secure the needed rights-of-way and tax dollars for freeway construction. The genesis of freeway construction in each of these cities dates back to the turn of the century as city planning departments included highways in their master plans. Houston, however, would construct the Gulf Freeway and thereby set the precedent for highway design and project management. Again, the Gulf Freeway was the first toll-free freeway constructed in the United States after World War II; its first 4.1-mile section opened on September 30, 1948.

The City of Houston was particularly effective in securing the Gulf Freeway. Mayor Holcombe secured the right-of-way of the Galveston-Houston Electric Railway Company back in 1939. The Galveston-Houston Interurban applied to the city to abandon its tracks within the city during that year, and Holcombe purchased that right-of-way for the city, which later provided part of the Gulf Freeway's route through the city as well as enhancing the reputation and popularity of the mayor. "I said then that we would someday build one of the finest highways in the country on this right of way," Holcombe recalled.[36] The planning commission followed the mayor's lead by requiring real estate developers to recognize the right-of-way requirements for this major thoroughfare in the platting of new subdivisions.[37]

The first urban freeway in Dallas was the Central Expressway. The idea for the Central Expressway, much like many urban freeways in the state, had its genesis at the turn of the century and was first manifested in city engineer George E. Kessler's 1910 city plan. Kessler envisioned a divided crosstown boulevard in the city. The construction of this road would require that the tracks of the Houston and Texas Central Railroad be removed from Central Avenue, because the roadway needed a minimum of a 200-foot right-of-way, a width many believed excessive.[38]

Kessler's design called for the creation of a 70-foot park-like median dividing the two 40-foot lanes for incoming and outgoing traffic flanked by a 25-foot green space with trees and sidewalks.[39] Needless to say, today's Central Expressway is a far cry from Kessler's original design, but the idea of a centralized crosstown road that specifically served as an exclusive facility for vehicular traffic was translated into the modern roadway. Ironically, the City of Dallas and the Texas highway department began reconstructing the Central Expressway in 1992 and included many features that attempted to "dress up" the freeway by incorporating designs and landscaping that would convert its previous utilitarian design to one that motorists would supposedly find more aesthetically pleasing.[40] One critic quipped that incorporation of these elements gave the freeway a "Disney-esque feel."[41]

The efforts of Dallas to secure the right-of-way for the expressway proved to be problematic when the Houston and Texas Central Railroad, and later Southern Pacific Railroad, resisted the idea of surrendering right-of-way. The railroad blazed the path of the future freeway when it built eleven miles of track through Dallas in 1872. After numerous meetings with city officials, the railroad agreed to abandon its tracks in exchange for $2 million in April 1923.[42] Though city officials had pledged to make the construction of the central boulevard a top priority, no work was done on the Central Expressway until the Texas highway department began survey work in 1937 and prepared cost estimates in 1939. Construction began in 1947, and the first two miles of the Central Expressway officially opened in 1949.[43]

Much like other urban freeways in the state, the number of cars on the Central Expressway quickly outstripped the roadway's capacity. The expressway's capacity was exceeded in 1971, when its 75,000 vehicles-per-day capacity could not handle all of the cars from Dallas and the suburbs of Richardson and Plano. Much like other projects in the state, expanding the capacity of the Central Expressway did not happen imme- diately, because the sheer cost of the expansion was not considered palat- able by state and local politicians, especially when the Dallas Area Rapid Transit bond failed in 1988. In sum, users of the Central Expressway had to live through twenty-eight years of traffic congestion before something was done about it. The Central Expressway remained congested until expansion and renovation began in 1990. The renovated road was designed to handle 240,000–270,000 cars per day. Revamping the expressway was complete by 1999. By 2005, some critics of the project noted that conges- tion would soon again reappear precisely because, as they argued, "Texans love their cars too much to allow a new freeway to go unclogged."[44]

The construction of the Gulf Freeway clearly signaled an endorsement of the automobile as Houston's primary mode of transportation and a dedication to the construction of automotive transportation facilities. But traffic here also quickly outpaced designed capacity, and the freeway did not remain a unique facility. Over time, engineer-manager William James Van London and his successors designed and oversaw the construction of a massive loop-and-spoke system of freeways for Houston.[45] The Gulf Free- way proved to be popular with the public and served to fuel their desire for more and more automotive transportation facilities. Nevertheless, Hous- tonians quickly discovered that the freeway's popularity made it one of the most congested freeways in Texas, bringing on a continual demand for more freeways. Interestingly enough, Houstonians also complained about the never-ending maintenance and repair of the existing freeway network while demanding the construction of new facilities. Traffic engineers and others were prompted by these complaints to study the facilities in an effort to understand urban traffic congestion and the facilities' role in it. This scenario was played out in every major urban center in Texas.

The Gulf Freeway's popularity can be easily seen in the rapid increase of the numbers of Houstonians who used it. The designer's original usage projections had been very low, since it had been designed before the war ended and was based upon prewar usage projections. But the size of postwar migration to Houston and the sheer number of automobiles that competed for space on the city's roads were outside of the scope of pre- war data. Many highway planners believed that the construction of urban freeways would reduce congestion.[46] This, however, was not the case. The

Texas Transportation Institute found that the initial opening of the Gulf Freeway and other urban freeways did offer the Houstonians a rather high level of mobility, but the size of the traffic loads quickly outstripped capacity, leading to congestion and longer peak periods.[47] This method of predicting postwar freeway capacities based on prewar data is also typical of other Texas cities.

In 1948 the *Houston Chronicle* reported that the capacity of the Gulf Freeway was 70,000–100,000 automobiles per day but that normal daily capacity was expected to be 25,000–30,000 "once sightseeing has sub-sided."[48] This projection was erroneous. Engineer-manager Van London reported that 80,000 automobiles per day were using the roadway in 1952, a number that "far exceeds the original estimates of the traffic flow expected." Then in 1955 he revealed that some 112,000 vehicles used the freeway daily and that the peak traffic load between Cullen Boulevard and Dowling Street reached 90,000 vehicles per day. By 1960 the roadway reached its saturation point, carrying more than 100,000 vehicles per day, a volume it had not been anticipated to reach until ten years later.[49] Although Van London did not live to see it, his roadway rapidly became the most congested freeway in the state, a title it held until 1971.[50]

Houston was not the only city to experience traffic congestion on its emerging network of freeways; the same scenario was repeated through-out the country. The major cities of Texas were all suffering from rising congestion even though they all were in the process of constructing free-way networks.[51] As the rest of the nation would find out as time passed, the promise of the freeway—unhindered mobility throughout the urban landscape—proved impossible, for there was simply not enough money to construct new roadways continually to handle the ever-increasing demand.[52] Urban highways had been seen as the only way to save America's urban centers from decay during the postwar era, but this turned out to be false. Congestion levels rose so high nationally that 60 percent of the existing federal highway system, some 660,000 miles of roads, was inade-quate to handle the demands put on it by the driving public by 1952.[53] The national postwar policy of the continual building and expansion of freeway networks failed to relieve congestion in America's urban centers.[54]

It is important to understand that this pro-freeway agenda was com-pletely devoid of concern for its environmental impact; its focus was on growth and expansion.[55] The shift from the war to peace also brought the construction of infrastructure and housing. Texas urban growth was severely constrained by the federal controls on housing materials, and the vast numbers of returning veterans and migrants to Texas's metropolitan areas helped to make the pro-growth and pro-freeway agenda popular.[56] Yet the immediate concerns with local infrastructure needs of a continually expanding population superseded any environmental concerns. By the time these environmental issues carried any political weight, Texas's cities were already sprawling urban masses.

As the promise of limitless mobility began to break, traffic engineers searched to find ways to understand and identify the causes of congestion. The Gulf Freeway became the most studied freeway on the face of the globe. Thomas E. Willier, Houston's traffic manager, began to study the freeway's effects on local street traffic in 1948, and the consulting engineering firm Norris and Elder examined the facility's impact on land values and land use at the request of the Bureau of Public Roads in 1951 and in 1956. These two studies set the groundwork for future freeway studies despite the fact that both of them contained no environmental impact component. They both showed an overriding concern with the tangible economic effects of the construction of expressway facilities in urban areas, and both were used as propaganda to encourage other urban centers to construct such facilities. As a result, the environmental problems created by an autocentric city were seemingly universal.

The construction of the Gulf Freeway has affected Houston and the region it serves tremendously. More important, this freeway served as the prototype for all urban freeways constructed in the state. Though many Houstonians are blind to this fact, the freeway has greatly influenced the built landscape of the city, defining the very nature of the people who live in Houston, Galveston, and the surrounding suburbs. The Gulf Freeway was the first step in the creation of Houston's modern built environment, which is often characterized by its freeways. The built environment of Houston was remade to accommodate the roadway as hundreds of structures were moved or demolished in the central business district alone—displacing, here and elsewhere, hundreds to thousands of African Americans and other minority or working-class Texans in the process and without fully compensating them with adequate replacement housing.[57]

Future highway planners in Houston and the other cities of Texas continued to replicate the Gulf Freeway throughout the region in hopes that it would have the same expected effects as the original. This action created sprawling urban forms and fostered automobile dependency and

congestion as travel distances increased. Texans had once seen the Gulf
Freeway as a panacea for all of the state's congestion problems, but as the
system aged and more commuters joined the ranks of the existing ones the
"deluxe superhighway" became just another roadway to wait on as traffic
slowed down.

Urban freeways do offer a convenient means for moving about the
region, but they also contribute to air pollution. Studies of air pollution
in the Los Angeles region have shown that automobiles are responsible
for 70–80 percent of the region's pollution, an estimate that likely holds
true for Texas as well.[58] Freeway systems, when not in a state of per-
petual congestion, actually reduce automobile emissions by speeding
traffic through an urban area. But the promise by many traffic engineers
across the nation, including Van London, that urban freeways would be
congestion-free attracted more vehicles to the freeway than engineering
and feasibility studies projected. Traffic engineers had discovered the
traffic-generating facet of urban freeways as far back as the 1950s, but this
was generally glossed over, as it was in Willier's 1949 *Economic Evaluation
of the Gulf Freeway.*[59] Such congestion created a voluminous amount of
auto emissions in the local air, which when combined with other pollutants
and sunlight results in a thick blanket of smog. Radial freeways such as the
Central Expressway, in other words, are particularly bad about generating
traffic congestion and, hence, smog.[60]

Houstonians, as well as many others from the surrounding towns
and suburbs, became increasingly automobile dependent in the postwar
era. The increased amount of suburban housing development served to
reinforce the necessity of a commuting lifestyle as the distance between
job and home continued to increase. The explosive suburban growth
that accompanied the Gulf Freeway and other roads gave Houston the
unwanted nickname of "the monster that ate East Texas," as wave after
wave of construction washed over the Gulf Coast region in tsunami fashion, leaving behind a dizzying array of suburban houses, shopping malls,
and commercial establishments. With the fostering of such growth to
accommodate new residents, commute times to the city increased as more
and more cars used the same roads from locations farther and farther out.
This again created traffic congestion and, especially when coupled with
the industries on the Ship Channel, created air so polluted that one could
see it by the early 1960s. Sadly, this state of affairs was replicated throughout the state as more and more miles of urban freeways were constructed.

The problem of air pollution is particularly acute in Houston because its industries and automobiles are both major contributors to the problem. Ironically, automobiles were thought to alleviate health problems that were specific to the city when they were first introduced, but the combination of ever-increasing numbers of automobiles and traffic congestion created a very different result.[61] The commuting lifestyle that generations of Houstonians adopted created massive amounts of automobile emissions—one of the biggest sources of environmental degradation in the Houston area and nationwide.[62]

Nationally speaking, California has been far more progressive in tackling this problem. It enacted statewide anti–auto pollution measures before the federal government imposed them on the rest of the country with the 1965 Motor Vehicle and Air Pollution Act and the 1970 Clean Air Act.[63] The air in Houston, however, remained thick with auto and industrial pollutants after these acts were passed, especially during "smog season"—the seven-month period beginning in April and ending in October. An Environmental Protection Agency study found that the elevated levels of ozone in the Houston area during this smog season are the result of calm winds and increased ozone production from a variety of sources.[64] Yet Houstonians, much like the rest of the nation, did not begin to consider air pollution a problem until the mid-1960s, and even at that point they were more concerned with traffic congestion. The failure of Houston, or any other Texas city for that matter, to examine any other kind of transportation alternative to the automobile is clear evidence of this trend. Furthermore, a 1978 survey discovered that, although a mass transit alternative was a widely popular idea, few residents were willing to pay for its implementation. The use of the automobile in the urban landscape of Texas was a deeply ingrained cultural habit by that point, and such habits are not easily broken.[65]

The success of the Gulf Freeway initiated a series of self-perpetuating cycles that have not only reinforced the automobile as the primary mode of transportation in the region but also defined trends in housing, shopping, and transportation planning. These cycles are not unique to Houston, by any means; one can easily find them in every major city in Texas and the nation. The first of these cycles was the direct result of improved access to land on the fringe, which was subdivided and rapidly developed into suburban housing. This movement to the suburbs was further encouraged by government loans to veterans and the availability of employment in the city and at the nearby Ship Channel industrial plants. This cycle was repeated as Houston's freeway network was expanded, extended into the hinterland, or improved. Although this expansion created affordable housing and a real estate and industrial boom, it also reinforced automobile dependency, contributed heavily to the number of automobiles on the rapidly congesting freeways, encouraged more freeway development, and added significantly to Houston's poor air quality. Many of these problems have either been glossed over or ignored until funds were no longer available to continue the cycle, or they reached a point where something had to be done about it. More often than not it would be the federal government that provided the impetus for change, either through federal law or by withholding money.

Texas's cities and suburbs, as in other Sunbelt areas, have paid a high price for the immediate gains associated with metropolises built for the automobile. The most obvious of these is dependence on the automobile to perform even the simplest of errands, now that few things are within walking distance. Everything, or so it seems, begins with a trip down the freeway. The shortsightedness of automobile dependency was made all too apparent during the evacuation preceding the arrival of Hurricane Rita in 2005. An estimated three million people poured onto the Gulf Freeway, creating massive gridlock within the region. Prompted by fear of the massive devastation caused in New Orleans during Hurricane Katrina, Houstonians, Galvestonians, Clear Lakers, and others packed up their cars for what will probably always be the largest exodus out of the area. Many people who did not live in the mandatory evacuation zones joined the swelling lines of traffic. The titanic convoy exhausted the supply of available gasoline as it moved slower than a turtle toward Dallas. The inability of the state to open both sides of Interstate 45 to outbound traffic created even worse congestion. This situation resulted in the deaths of an estimated sixty people, twenty-three of whom died when their bus exploded outside Dallas. Though it is debatable how much of this was directly attributable to the fear of another Katrina, the result shows that massive evacuation is at the very least difficult to accomplish.[66]

As the oldest freeway in Texas, the Gulf Freeway symbolizes many of the hopes that Texans and Americans had after the war—a quiet house in the suburbs, congestion-free mobility, and a good-paying job. As time progressed, the freeway also came to symbolize the aggravations of modern city life—gridlock, road rage, congestion, sprawl, and air pollution. This evolutional duality is not specific to the freeways of Texas. Many other roads were constructed along similar guidelines and design specifications during the construction of the interstate highway system.

The construction of urban freeways in Texas during the postwar era illustrates the dream of limitless and congestion-free mobility being largely deferred as the realities of overtaxed roads, limited funding, and a degraded environment become increasingly apparent to even the most skeptical observer. The construction of these urban freeway networks provided the Lone Star State with a boom in suburban construction, but it also fostered a seemingly cultural dependence on the automobile. Because no real alternatives to the automobile were ever considered by anyone on any level of government or by the citizens themselves, these problems associated with autocentricity will continue to play a major role in Texas transportation planning.

NOTES

1. Though it is not the focus of this chapter, it bears mentioning that transportation facilities also allow commuting by central city residents to and from the suburbs. Rarely do we think of this "reverse commuting" as an aspect of suburban transportation; it remains a largely unexplored topic in urban history. Too, it is becoming more common for suburbanites to commute from one suburb to the next.

2. Peter C. Papademetriou, *Transportation and Urban Development in Houston, 1830–1980* (Houston: Metropolitan Transit Authority of Harris County, 1982), 42–43.

3. Texas Transportation Institute, *A Mass Transportation Concept for Metropolitan Houston: Immediate Action, Flexibility, Economy* (College Station: Texas Transportation Institute, 1970), 6.

4. Richard F. Selcer, *Fort Worth: A Texas Original* (Austin: Texas State Historical Association, 2004), 73, 76–77.

5. Char Miller, *Deep in the Heart of San Antonio: Land and Life on South Texas* (San Antonio, TX: Trinity University Press, 2004), 11–12, 99–111.

6. Bret Wallach, "Ambidextrous Dallas," *Geographical Review* 99, no. 4 (October 2009): 460.

7. Perhaps the best example of this is Houston's annexation of the master planned community of Kingwood in 1995. This not only resulted in a failed protest but also currently holds the record for the single largest annexation in the state of Texas.

8. The planning department had been disbanded in 1927 after an attempt to implement zoning failed. "History of the Planning Department and Planning Commission," www.houstontx.gov/planning/AboutPD/pd_history.html (accessed February 2019).

9. Robert D. Thomas and Richard W. Murray, *Progrowth Politics: Change and Governance in Houston* (Berkeley, CA: IGS Press, 1991), 11; Amy L. Bacon, "From West Ranch to Space City: A History of Houston's Growth Revealed through the Development of Clear Lake," Master's thesis, University of Houston, 1996, 10.

10. "Houston Largest City in South," *Houston Chronicle*, January 16, 1950, Progress ed., 14.

11. "Persons living in suburban areas are spending hours daily getting to and from work or market. The average city street varying in width from 40 to 80 feet in width, cannot carry the traffic." "$2,000,000 per Mile Gulf Freeway Is the 'Cheapest Highway Ever Built,'" *Houston Chronicle*, January 20, 1952, sec. F, 9.

12. "If Only There Were More," *Houston Post*, December 31, 1948, sec. 2, Parade of Progress, 1.

13. Papademetriou, *Transportation and Urban Development*, 48.

14. Don E. Carleton, "Oscar F. Holcombe Collection," *Houston Review* 2 (Fall 1979): 136.

15. "Cars Here Increase Faster Than People; Street Plan Is Pushed," *Houston Chronicle*, January 20, 1952, sec. F, 2.

16. Robert H. Talbert, *Cowtown-Metropolis: A Case Study of a City's Growth and Structure* (Fort Worth: Texas Christian University, 1956), 18, 21.

17. There is some debate over whether Dallas is truly a landlocked city.

18. Rudolph Bush and Anne Miller, "Cities Reshaped through Annexation: Houston and Austin Continue to Grow; Dallas Landlocked," *San Antonio Express*, July 21, 1999, 1H.

19. United States Statutes at Large Containing the Laws and Concurrent Resolutions Enacted During the Second Session of the Seventy-Eighth Congress of the United States of America, 1944, and Proclamations, Treaties, and Other International Agreements Other Than Treaties, Complied, Edited, Indexed, and Published by Authority of Law Under the Direction of the Secretary of State, Volume 58, Part 1, Public Laws (Washington, DC: Government Printing Office, 1945), 838.

20. "Texas Roads and Highways" (Austin: Texas Legislative Council, October 1952, no. 52–3), 26.

21. Commissioner Thomas H. Macdonald of the Bureau of Public Roads, Washington, D.C., to Division Engineer J. A. Elliott of the Bureau of Public Roads, Fort Worth, Texas, August 27, 1947, National Archives, College Park, MD, Files of the United States Bureau of Public Roads (hereafter NA-BPR), RG 30, "Records of Division Six, May–August 1947," Box 1779, Folder 1.

22. Edward Weiner, *Urban Transportation Planning in the United States: An Historical Overview* (Westport, CT: Praeger, 1999), 17.

23. Commissioner Thomas H. MacDonald of the Bureau of Public Roads, Washington, D.C., to The Honorable Wright Pateman of Texas, Washington, D.C., January 26, 1949, RG 30, "Records of Correspondence FAS-Texas, January-June 1949," Box 2985–86, folder 2, 2–3, NA-BPR.

24. "Texas Roads and Highways," 79–80.

25. Weiner, *Urban Transportation Planning*, 27.

26. "Texas Roads and Highways," 24, 95.

27. Richard Morehead, *Dewitt C. Greer: King of the Highway Builders* (Austin, TX: Eakin Press, 1984), 46.

28. Remarks of division engineer J. A. Elliott at the opening session of the Division Six 1949 Bureau of Public Roads Administration meeting, RG 30, "Records of Division Six, January–March 1949," Box 1782, folder 3, 6, NA-BPR.

29. "$2,000,000 per Mile Gulf Freeway," sec. F, 9.

30. "Texas Roads and Highways," 25.

31. "Early in World War II, it was felt that all possible energies should be devoted to national war production. To place this new policy in effect, the national government suspended federal aid and discouraged all except vitally needed programs. These moves virtually suspended highway construction from 1942 through 1945." "Texas Roads and Highways," 94–95.

32. "Texas Roads and Highways," 24, 95.

33. Unknown to H. E. Hilts, Washington, D.C., February 11, 1949, RG 30, "Records of Correspondence FAS-Texas, January–June 1949," Box 2985–86, folder 2, NA-BPR.

34. D. W. Loutzenheiser, Washington, D.C., to J. Barnett, Washington, D.C., 12 December 1949, RG 30, "Records of Correspondence FAS-Texas, July–December 1949," Box 2985–86, folder 1, 1, NA-BPR.

35. Remarks of division engineer J. A. Elliott, 14.

36. "Opening of Our Splendid Freeway," *Houston Chronicle*, August 3, 1952, sec. E, 2; "Mayor Throws Switch to Light Superhighway; Houston-Galveston Expressway, Will Be First of Kind in Texas," *Houston Chronicle*, October 1, 1948.

37. Eugene Maier, Cooper McEachern, and Ralph S. Ellifrit, "Preliminary Report: Freeway Phase: Houston Metropolitan Transportation and Transit Study" (Houston: The City, 1961), copy located in the Texas Room, Houston Public Library, Houston, TX, 1.

38. Louis P. Head, "Kessler Proposed Boulevard out of Central Avenue," *Dallas Morning News*, December 30, 1924.

39. Head, "Kessler Proposed Boulevard."

40. David Dillon, "Central Expressway's Road Warriors," *Dallas Morning News*, November 10, 1987.

41. Chris Kelley, "A Carefree Tour of North Central Expressway in '98," *Dallas Morning News*, June 3, 1990.

42. The Southern Pacific surrendered this right-of-way despite the fact that it did so at a loss. See Head, "Kessler Proposed Boulevard."

43. "Central Expressway's Path through History," *Dallas Morning News*, December 5, 1955.

44. "Central Expressway's Path"; Henry Tatum, "Central Expressway," *Dallas Morning News*, December 1, 1999.

45. The "loop-and-spoke" design of urban freeway systems was a nationally typical design, thought to retard sprawl and revitalize deteriorating central business districts by speeding the flow of traffic in and around a built-up urban area to provide access to the central city. Mark H. Rose, *Interstate: Express Highway Politics, 1939–1989* (Knoxville: University of Tennessee Press, 1990), 60–61.

46. Rose, *Interstate*, 59.

47. Texas Transportation Institute, "Houston Corridor Study: Final Report" (College Station: Texas Transportation Institute, 1979), 3.

48. "Mayor Throws Switch."

49. "New Freeway Change-Over for Houston," *Houston Chronicle*, August 1, 1952, sec. C, 2; "Freeway Proves It's a Motorist's Best Friend," *Houston*, January 1955, 26; "100-mph Zone Ahead," *Houston*, November 1960, 44.

50. "Gulf Freeway No Longer Most Congested," *Houston Post*, February 7, 1982, sec. D, 2.

51. "Texas Roads and Highways," 48.

52. Kenneth T. Jackson, *Crabgrass Frontier* (New York: Oxford University Press, 1985), 270.

53. Richard O. Davies, *The Age of Asphalt: The Automobile, the Freeway, and the Condition of Metropolitan America*, ed. Harold M. Hyman (New York: J. B. Lippincott, 1975), 14.

54. Roadbuilding policy in postwar America considered the expansion of existing roads and the construction of new ones as the sole solution to the urban congestion policy. Charles O. Meiburg, "An Economic Analysis of Highway Services," *Quarterly Journal of Economics* 77, no. 4 (November 1963): 648.

55. This was also the agenda of the Galveston-Houston Electric Railway, which actively promoted the creation of suburbs such as Park Place. Contemporary newspaper accounts of the interurban's opening stated in December 1911, "The first and perhaps the most important effect of the interurban will be the settling up of the vast section of country between Galveston and Houston . . . capable of supporting several thousand population. The interurban will afford quick and easy transportation to these people; they will dispose of their garden produce and fruit in Galveston and Houston, and in these cities will do their trading, thus bringing added prosperity." Herb Woods, *Galveston-Houston Electric Railway* (Glendale, CA: Interurbans Special No. 22, 1976), 21.

56. See Paul Alejandro Levengood, "For the Duration and Beyond: World War II and the Creation of Modern Houston Texas," Ph.D. dissertation, Rice University, 1999.

57. Eric Avila, *The Folklore of the Freeway: Race and Revolt in the Modernist City* (Minneapolis: University of Minnesota Press, 2014).

58. Steven Nadis, James J. MacKenzie, and Laura Ost, *Car Trouble* (Boston: Beacon Press, 1993), xvi.

59. Gary T. Schwartz, "Urban Freeways and the Interstate System," *Southern California Law Review* 49 (1976): 490.

60. George W. Hilton, "Transport Technology and the Urban Pattern," *Journal of Contemporary History* 4, no. 3, Urbanism (July 1969): 132.

61. James J. Flink, "Three Stages of American Automobile Consciousness," *American Quarterly* 24, no. 4 (October 1972): 455–56.

62. Joel L. Naroff and Bart David Ostro, "The Impact of Decentralization on the Journey-to-Work and Pollution," *Economic Geography* 56, no. 1 (January 1980): 63.

63. Rudi Volti, *Cars and Culture: The Life Story of a Technology* (Westport, CT: Greenwood Press, 2004), 120–21.

64. Marilyn Davis and John Trijonis, *Historical Emissions and Ozone Trends in the Houston Area* (Washington, DC: United States Environmental Protection Agency, EPA-600/S3–81–030, July 1981), 2–3.

65. Flink, "Three Stages," 468.

66. Liz Austin, "Task Force Aims to Prevent Another Hurricane Evacuation Nightmare," Associated Press State and Local Wire, December 22, 2005.

{ 6 }

RISING UP

THE ASCENT OF IRVING TO SUPER-SUBURB
AND HOME OF THE DALLAS COWBOYS

PHILIP G. POPE

Until 1969, Irving, Texas, lacked a distinct narrative identity. For nearly half a century Irving had been subordinate to the older, more developed Dallas. Even though Irving had experienced substantial growth since the 1940s, and especially since the mid-1950s, it still lacked something. It lacked a landmark, a symbol; Irving was not readily identifiable with any-thing. Even into the 1960s, Irving was still dependent upon, and associated with, Dallas, not so much its neighbor to the east but rather its controlling overlord whose grasp it could not seem to break from. Whispers of a change in Irving's subordinate status began to be heard in 1967, and by 1969 the whispers had evolved into sustained conversation. By the 1970s, Irving possessed its own identity, thanks in large part to Texas Stadium, home of the Dallas Cowboys.

Texas Stadium did not cause Irving to suddenly become an independent city. As with other Texan communities explored in this volume, for many years Irving had been annexing land, acquiring new businesses, growing in population, and improving its infrastructure. Irving was seen, though, as simply one of many Dallas suburbs and, as such, grew with Dallas. The acquisition of Texas Stadium, though, showed Irving to be a player in the North Texas metropolitan cityscape. Irving was able to show that it could stand on its own and be heard, even in competing against Dallas. Equally important, Irving gained its own identity and became recognizable on the national map and in the national consciousness.

Prior to the construction of Texas Stadium, Irving was not widely identifiable on the national map. It was only in 1965, in fact, that Rand-Mc-Nally decided to put Irving on its maps. The location of Irving was also seemingly ignored by some in the state. At the Beltline Road exit of the Dallas–Fort Worth Turnpike in 1966, there was no sign indicating that the exit was for Irving. The Texas Turnpike Authority even refused a City of Irving request to place Irving on the sign, although Grand Prairie was already so represented.[1]

During their travels, Irving Chamber of Commerce members were regularly asked where Irving was located. While traveling to major cities such as Boston and Chicago as part of the East Texas Chamber of Commerce, members of the large Irving delegation constantly had to explain where Irving was. These Irvingites were also asked the same question in their own state. When visiting the capital, Austin residents wondered where Irving was located. Even a state representative from Waco, a mere hundred miles south of Irving, was unsure of its location.[2]

It was not as if Irving had attempted to isolate itself from the rest of the world; it had in fact tried to make others aware of its location and amenities. Indeed, the chamber of commerce was quite active in spreading the word about Irving. The chamber liked to boast that Irving had grown over forty-fold in less than two decades, transforming from a "small, scattered community" into a "fast-moving city." At the 1966 Texas state fair, the chamber of commerce, along with the Jaycees, sponsored a booth to boast about Irving.[3]

Even with the growth Irving experienced, it failed to separate itself from the other suburbs around the Dallas–Fort Worth Metroplex. Whether Irving was an independent community or simply a bedroom community was debatable. It experienced sustained growth in the 1950s, each year gaining roughly 5,500 new residents, the majority of whom commuted daily to work in Dallas. Despite efforts to separate it from Dallas, Irving's identity was still not its own. To combat this, the chamber of commerce pushed through projects that might provide Irving its own identity, the two major examples being the renovation of downtown and work on the Trinity River Canal project.[4]

These projects failed initially to bring much more than disorder to the face of Irving. Chamber of commerce member Bill Stevens drove around the city for about an hour in 1966 to see what construction activity was taking place. He was forced to take multiple detours on his trip, due to the closure of existing roads and the construction of new ones. According to Stevens, a certain detour arrow seemed "to lead nowhere." He hoped that no one else would similarly get as confused as he. His experience was symbolic of the growth Irving experienced. Changes were occurring as the city continued to grow. The problem was that there was simply no order or direction to this growth and, hence, no order or direction concerning the city itself.[5]

One could partially attribute Irving's lack of goal attainment to apathy among its citizens. Irving residents seemingly did not care about the city's development; they were content to have Dallas leaders dictate Irving's future. When the Dallas–Fort Worth Regional Airport was proposed, Dallas mayor Erik Jonsson addressed the citizens of Irving as if they were subordinate to, and not independent from, Dallas. Jonsson told the citizens of Irving that such an airport was needed for the area, and even if those in the surrounding counties did not vote for the airport, "the two cities," meaning Dallas and Fort Worth, would. This was but one of many occasions on which Mayor Jonsson would come to Irving and address its people as subordinates of Dallas.[6]

Out of Irving's total population of roughly 84,000 in 1966, only 19,000 were registered to vote; of those 19,000, a mere 2,500 had exercised their right in the previous election. Democracy advocate Jack Harkrider addressed the Irving Chamber of Women on the lack of voter turnout. In the cold war–influenced language of the era, he warned that, on the local level, this apathy could reverse the positive growth trend that Irving experienced. On a larger scale, he warned that such apathy would present communists with an opportunity to dissolve American democracy.[7]

Regardless of whether Irvingites were concerned with a communist takeover, voter turnout did begin to increase, with most voting for an improved and progressive city. In 1967 the school board and city council, among other groups, urged voting in favor of the new airport. These groups saw the airport as a prime opportunity to bring new businesses to Irving, and thus increase the tax base, which would in turn give Irving more flexibility. The first vote saw both Dallas and Tarrant counties reject the new airport, while Irving stood alongside the bicounty metropolitan giants Dallas and Fort Worth in support of the airport. The citizens of Irving were beginning to look to the future, and though there were still hurdles ahead Irvingites were beginning to vote for progress. It would take a sports franchise from Dallas to prove just how committed Irving residents were to progress.[8]

Since their inception in 1960, the Dallas Cowboys had played their home football games in the Cotton Bowl, part of Dallas's Fair Park. Built in 1930, the 46,000-seat Cotton Bowl hosted college football games, plays, and speeches, among other events. With the growing popularity of the football games, the City of Dallas in 1947 sold bonds to renovate the stadium and expand its seating capacity to nearly 68,000. In 1949 the stadium was renovated again, and seating expanded to almost 76,000. The early 1950s brought increased usage to the Cotton Bowl: in 1950 the Dallas Eagles baseball team of the Texas League played there; and, in 1952, the National Football League's Dallas Texans played in the Bowl for one notoriously horrible season (they went 1–11). Then, in 1960, the Cotton Bowl became the home of the NFL's Cowboys and the American Football League's Dallas Texans (1960–63; they then moved to Kansas City and became the Chiefs).[9]

The Cotton Bowl provided Dallas a venue in which sports teams could perform their magic. By the 1960s, though, despite the previous expansions, the Cotton Bowl's magic and charm had diminished. Although the stadium contained nearly 76,000 seats, barely half of them were located between the end zones, which is where the majority of the premier spectator seating is located. In addition to the undesirable location of the seats, they had a reputation of being uncomfortable. In 1967, following seven losing seasons, the Dallas Cowboys, first led by Dandy Don Meredith, started winning, producing crowds that overflowed the 40,000 good seats. Clearly something needed to be done to accommodate the increasing number of fans.[10]

Even though $2.5 million was allocated for the expansion and improvement of the Cotton Bowl, many doubted the expensive facelift would be sufficient to improve the stadium enough to host an emerging and dominant NFL franchise. To make the Cotton Bowl a desirable place to play, the city would also have to improve the neighborhood, for the area surrounding the Cotton Bowl and Fair Park was noted for its poor parking conditions, traffic jams, and bad neighborhoods. After games there were regular incidents of assaults, robberies, vandalism, and isolated instances of stabbings and shootings in the immediate vicinity. In view of the area surrounding Fair Park, the Cotton Bowl's renovation project was dubbed "Operation Rathole."[11]

Cowboys owner Clint Murchison Jr. was aware of the problems associated with the Cotton Bowl and had been looking into a new stadium since 1963. Murchison wanted to build a new football-only stadium in downtown Dallas. This, he insisted, would not only keep his team in Dallas but also assist in revitalizing downtown, which was essentially empty on the weekends. Neither Dallas mayor Erik Jonsson, mayor pro-tem Frank Hoke, nor the Dallas city council seemed interested in Murchison's desire for a new stadium. The city council, according to future Irving mayor Dan Matkin, had a "love affair" with Fair Park and they were determined to save it, fearing that a stadium built elsewhere would lead it further into decline. Too, Jonsson, though not willing to allow a new stadium, did not even see the need to improve the Cotton Bowl.[12]

Rebuffed by the leaders in Dallas, Murchison began to look outside of Dallas for a new stadium site. His first thought was neighboring Irving, since the suburb could provide a convenient location for most fans to attend games, although leaders in South Dallas would later complain that Irving was too far for the fans there to drive. Murchison instructed his associate, Max Thomas, to contact Irving mayor Lynn Brown regarding the possibility of relocating the Cowboys to Irving. Neither Brown nor select members of the Irving city council were interested in the plan and simply refused to meet with Murchison.[13]

In addition to Brown and city council members, Murchison had to be concerned about the residents of Irving. Though voter turnout had been increasing in the 1960s, and citizens were voting for more progressive legislation, Murchison knew that voters might still be a problem. Irving voters seemed more concerned with the tax effect of a new stadium than with anything else. Residents cited Houston's recently completed Astro-dome as a project that caused an increase in taxes for that city's residents. Some Irvingites, though, conceded that, although the Astrodome did cost the taxpayers, it nonetheless benefited the entire community; it was her-alded as the "Eighth Wonder of the World" and brought much attention to Houston, which was showing its progressive face to the world.[14] Still, many Irvingites, especially the older ones, remained opposed to the idea of taxpayer money contributing to a new stadium. Indeed, because of these factors and Irving's track record on public projects, most Irving residents simply did not believe that relocation to Irving would occur.[15]

Frustrated, Murchison expanded his search farther from Dallas. According to former Irving planning and zoning commission member Melvin Schuler, Murchison considered distant Oklahoma City as a possible new site, though he wanted to keep the team as close to Dallas as he could. Arlington appeared to be the next best alternative, for it was a growing city that was accessible from many directions. After traveling to Arlington from Dallas, though, Murchison decided the drive was simply too long. He had not given up on Irving and decided to try again, this time meeting with people other than Mayor Brown.[16]

Murchison became convinced that Irving was the ideal location for his team. The Cowboys were the football team not merely of Dallas but of the entire metroplex—and Irving was located at the center of the metroplex's ever-growing population. In fact, the site that Murchison eyed was closer to greater Dallas's population center than was Fair Park. The estimated commute times to the proposed stadium were very reasonable from most parts of the metroplex, requiring only twelve minutes from Richardson or Carrollton, thirteen minutes from Grand Prairie or Oak Cliff, eighteen minutes from Arlington, nineteen minutes from East Dallas, and only thirty minutes from Fort Worth.[17]

In December 1966, Murchison arranged a meeting with fifteen to twenty progressive leaders of Irving at the Dallas Gun Club. In attendance were Schuler and council member Robert Power, along with various members of the chamber of commerce and other community leaders. Murchison presented his plan of a stadium in the "teardrop area," where the future Loops 12, 114, and 183 would meet and form a triangular area of roughly 100 acres, and where at present there was located a junk yard, trailer park, and nursery. Irving, not having a substantial tax base, lacked the finances to secure all of the proposed land. Realizing the problems Irving may have in acquiring the land, Murchison agreed to purchase the entire area, and Irving could buy it at cost through municipal bonds. If the stadium deal did not go through, Schuler and Thomas could move ahead with their plans for a commercial development on the land.[18]

The main problem Murchison saw was the lack of cooperation with Mayor Brown. Murchison was all too experienced in dealing with an uncooperative mayor in Erik Jonsson and scarcely wanted another such relationship. At the end of the meeting at the Gun Club, Murchison instructed Schuler and Power to "get you a new mayor." Being that 1967 was an election year, a decision had to be made quickly. Schuler initially decided to run, but the fact that he was going through a divorce discouraged him from such an effort. A week after the Gun Club meeting, Power decided to run for mayor against Brown.[19]

The 1967 Irving mayoral campaign would turn out to be a hotly contested, down-to-the-wire election. Eight-year incumbent mayor Lynn Brown, although unpopular with Murchison, remained popular with the more established voters of Irving. The previous election year, 1965, saw Brown run uncontested, and he became the first mayor of Irving to serve a fourth term. Power's decision to run for mayor in 1967 had less to do with getting the Cowboys and more to do with general ideology and philosophy, however. The previous twenty years saw Irving's population soar, and many of the newer residents felt as though Mayor Brown did not represent them as much as he did the older, more established residents. Brown's father had been one of the city founders, so he naturally had a connection with the Irving establishment. Power's main objective was to bring Irving's newer residents into the fold, to encourage their participation in city issues. Although the relocation of the Cowboys was not a main campaign issue, a vote for Power was, in essence, a vote to move in a cautious yet progressive manner in courting the team.[20]

In the initial election results of 1967, none of the three candidates garnered the necessary 50 percent of the vote, but Brown received more than Power. In the subsequent runoff election held three weeks later that saw record voter turnout, Power managed to edge out Brown. Newer Irvingites made their voices heard for a more progressive Irving that would take advantage of the city's vast opportunities for growth. Now Murchison had his man in office, and the path to Irving became a little clearer.[21]

Still feeling a bond with Dallas, and perhaps using Irving's progressive new leadership as leverage, Murchison met again with Mayor Jonsson, members of the Dallas city council, and state fair officials. Insiders claimed the meeting became hostile between Jonsson and Murchison, though in the press Murchison referred to it simply as "unrewarding." Dallas architect and city plan commission chairman Ross Ramsay still hoped that the Cowboys would stay in Dallas, and he proposed that the $2.5 million in bonds still be put toward renovating the Cotton Bowl, but only as an interim measure. Ramsay proposed a new football stadium and sports center to be constructed in downtown Dallas, near Memorial Auditorium. The hope Ramsay had, though, was just that. It seemed as though Dallas politics would simply not accommodate the Cowboys. The council insisted that Fair Park was the only option, and Mayor Jonsson remained staunchly against any plan for a downtown stadium, hoping instead for a downtown development that would provide year-round activity. The ball was then Irving's to run with.[22]

In April 1967, Irving voters cast their votes again, this time concerning forty-year municipal bonds for the proposed Texas Stadium. By selling these bonds, Irving would not have to resort to increasing the taxes of its citizens. Bond purchasers would receive the rights to seating at Texas Stadium, whose revenues would in turn pay for the matured bonds, which the bond holders could, at maturity, redeem for a small rate of return. Although the vote passed, the state attorney general refused Irving permission to issue the bonds.[23]

The next year saw a frenzy of activity in regard to the second election for the bonds; 1968 also saw more sparring between Murchison and Jonsson. Writing a guest column in the *Dallas Times Herald*, Murchison began by stating that he would "like to scattershot," but that he "used his last load arguing for that downtown stadium." After some filler material in which he rambled about the Cowboys over the past few years, Murchison closed by admitting that he had "already picked out my Christmas present for this year. I'd tell you, but it might put Erik on the spot." Whether or not Murchison still thought a downtown stadium possible is debatable; what remains clear is that Murchison, not Jonsson, held the upper hand.[24]

In January 1969 the bonds were issued, and subsequently the stadium issue was resolved. In the biggest voter turnout concerning a municipal election, Irving voters approved the bonds by a vote of two to one. In total, $31 million worth of bonds were sold. Murchison made up the difference between the value of the bonds and what he paid for the land. He then presented a deed for the land to Mayor Robert Power, who accepted on behalf of the City of Irving. Irving citizens had demonstrated that they were civic-minded, and that they could compete with Dallas and win.[25]

Irving was coming of age and gaining recognition. A couple of weeks after the bond election, Murchison unveiled a picture titled "Your Bond Money in Action . . . Texas Stadium." Pictured was the Roman Coliseum, albeit with a Cowboy star under one of the arches. In essence, Murchison was implying that Texas Stadium would be a magnificent sports stadium that would command attention. Chicago citizens were impressed by the whole process and the coup that Irving was pulling off, prompting *Chicago American* columnist Harry Sheer to ask, in regards to Wrigley and Soldier Field, "What has Texas got that Chicago hasn't?" The Cotton Bowl was a newer stadium than both Wrigley and Soldier Field. Those Chicagoans wishing for newer sports stadiums were apparently taken aback by what a suburb like Irving was able to do that Chicago had not yet done.[26]

Dallas was also taking notice of its neighbor. *Dallas Morning News* columnist Mike Kingston declared that Irving was "doing more than sticking their thumb in the Dallas pie and pulling out a plum." Indeed, Dallas was taking notice of Irving in a manner that it had not previously done. Kingston went on to state that "the spirit and perseverance displayed by Irving leaders in development of the project is not unlike that which built Dallas itself." Irving was coming into its own. It had gained recognition as a city that could do what it set its mind to, no matter the competition.[27]

Even with the passage of the bonds, there was still much to do. Irving had gained recognition, but the goal was to maintain this recognition and demonstrate that the city had much to offer. Though clearly not the only item of interest or value in Irving, Texas Stadium would command attention—attention that Irving hoped would showcase the emerging super-suburb, or what Robert B. Fairbanks labels a "suburban city" in his chapter for this volume. The Irving Chamber of Commerce sent out invitations to the groundbreaking of Texas Stadium, the footer of which read, "The Fastest Growing City in Texas." At the groundbreaking ceremony, Reverend Robin Moffatt gave the invocation, asking that Texas Stadium "be the place where the Cowboys bring additional pride to themselves and to this city." He also "called upon the All Mighty to make Texas Stadium a spot . . . everyone will look upon as one of the landmarks of Irving."[28]

The sense of identity that Texas Stadium brought to Irving turned out to be of supreme importance. Irving was previously viewed as simply a subordinate suburb of Dallas. Prior to Texas Stadium, the most recognizable landmark in Irving was the Braniff Memorial Tower, which though located in Irving was part of the University of Dallas campus, so it was still connected more with Dallas than with Irving. Texas Stadium was undeniably Irving's, and at roughly the same height as the Braniff Tower the stadium would be visible for miles around.[29]

Texas Stadium, when completed, would be large enough that the so-called Eighth Wonder of the World, the Houston Astrodome, would be able to fit inside, with room to spare. The open-dome top of Texas Stadium would be visible from downtown Dallas and most of the freeways in northwest Dallas County. Irving was clearly gaining a landmark. In 1971 the *Irving Daily News* unveiled its new emblem, which consisted of the outline of the state of Texas, a star to place Irving, the Braniff Tower, an airplane, and, pictured prominently and in detail, Texas Stadium.[30]

Irving also lobbied to acquire more sports teams. Texas Stadium Corporation talked with North Texas State University about playing at least one game a year at Texas Stadium. In another blow to Dallas and Fair Park,

Southern Methodist University began discussing the possibility of playing its home games at Texas Stadium instead of the aging Cotton Bowl. The acquisition of the Dallas Cowboys may have been Irving's first major jab at Dallas, but it would not be the last.[31]

Irving was turning into a destination, and not simply a place between Dallas and Fort Worth. As reviewed by Tom McKinney in his chapter for this volume, the construction, expansion, widening, and improvement to several area highways in the 1960s made places such as Irving easier to get to. Once there, Irving, by 1971, provided its 100,000-plus residents with twenty schools, more than one hundred churches and civic clubs, and seventeen parks. Irving was Texas's fastest-growing city in the 1960s. That decade saw the city grow by over 111 percent. At the beginning of the decade, Irving had been Texas's twenty-third-largest city, but by the end of the decade it had moved up eleven notches to become the twelfth-largest.[32]

The housing of Irving residents demonstrated that the city was becoming more diverse. Prior to the postwar population boom, Irving was a town whose residents lived in single-family houses, and two decades later the city had added newly built apartment complexes. As also discussed in Fairbanks's chapter in this volume, Irving's newer residents were giving the city more of an urban look, a departure from the establishment's suburban bedroom community appearance.[33]

Likewise, the increased population created a demand for better transportation infrastructure. In 1967 a portion of Highway 114 dubbed "Death Alley" had its speed reduced from 70 to 60 mph. By the mid-1960s the rapid growth had caught up to the city, and city leaders prioritized the improvements and extensions of Highways 183 and 356, Beltline, Northgate, MacArthur, and Spur 635. With the increase of population, commerce, and, hence, traffic, these improvements were necessary to keep pace with the city.[34]

Within a few years, more road improvements were planned, and there became a need for a road directly connecting Irving with Dallas; Fort Worth also saw the need for a direct link with Irving. In 1971 a toll road that would touch Irving and run parallel to the Dallas–Fort Worth Turnpike was planned. This road, called the Trinity Route, was not important in that it would connect Dallas and Fort Worth, for they were already connected. The Trinity Route was important because it included Irving, a super-suburb of growing importance.[35]

Texas Stadium became an advertising tool for Irving. According to Dan Matkin, before the coming of the stadium businesses would "ask where Irving was. Now they know." Indeed, Irving experienced more commercial growth after the decision to build Texas Stadium there. The value of building permits for the first two months in 1969 exceeded those from the same period of 1968 by 22 percent. That same year, the Irving Chamber of Commerce traveled to Washington, D.C., in an attempt to gain more business for Irving. On the trip, the group cited Texas Stadium as an example of the city's willingness to work with business. Murchison, a Dallasite, said he would advise anybody to invest in Irving.[36]

A number of regional and national companies relocated to, or built new facilities in, Irving in the late 1960s and early 1970s. For example, Sears chose Irving's new mall to locate another one of its stores. Carson Pirie Scott entered the Southwest market with its new distribution center in Irving. Time-DC, the nation's third-largest trucking company, opened a new terminal in Irving. The addition of Carson Pirie Scott and Time-DC demonstrated Irving's prime location. Not only was it convenient to Dallas, Fort Worth, and the new regional airport, but its improved highway system made it convenient to transport goods to regional cities like Houston and Oklahoma City.[37]

After spending twenty-three years in Dallas, Storey Electric Company relocated to Irving as well. Not only was Irving gaining new businesses from across the country, but it was also taking them away from Dallas. Many of these businesses may not have moved to Irving if not for Texas Stadium. The stadium gave Irving a sense of place, much like Disneyland did for Anaheim, California. Disneyland also assisted Anaheim in gaining businesses and recognition. Similarly, Stanford's Industrial Park enabled Palo Alto to gain name recognition and take business away from San Francisco. People and businesses began readily to realize where Irving was located.[38]

In an effort to accommodate its increasing business and population, Irving—like other Texan communities reviewed by Robert B. Fairbanks, Andrew Busch, and Andrew C. Baker in this volume—became more aggressive in the annexation of land. The first substantial acquisition of land occurred in 1953 when Irving expanded its boundaries to encompass eighteen square miles. Between 1950 and 1960, Irving's area increased by an astonishing 1,750 percent, then it roughly doubled between 1962 and 1974. Much of the annexation was done at the request of those living in the area to be annexed. Some, like Ben Carpenter, wished not to be annexed, however. As a compromise, Irving annexed only a protective strip around Carpenter's land, so as not to allow another municipality to annex that same land. Other annexations were induced because other municipalities, such as Grand Prairie, similarly extended their own protective strips around areas where future growth was predicted. Regardless, Irving was growing, and it had room to do so.[39]

Dallas leaders saw the increasing power of Irving and made attempts to consider Irving in their planning. The *Irving Daily News* reported in 1969 that an unnamed Dallasite asked for the formation of a "Super Chamber of Commerce," which would bring together the interests of Dallas, Fort Worth, and "the surrounding suburbs." There is little doubt that, aside from Fort Worth, and perhaps Garland and Arlington, the major goal was to get Irving on board, for Irving was quickly gaining greater influence as well as a broader tax base. Whereas the Mid-Cities contingent of Grapevine, Bedford, Euless, and North Richland Hills could legitimately be referred to as suburbs, Irving had evolved into a super-suburb of increased magnitude.[40]

In 1970 the Irving Chamber of Commerce was asked to visit Dallas to discuss the plans for the Dallas–Fort Worth Regional Airport. It is doubtful that, prior to acquisition of the Cowboys and the construction of Texas Stadium, Dallas would have conferred with the leaders of Irving. However much Dallas disliked the thought, though, Irving was no longer a subordinate suburb dependent on Dallas. The new airport was expected to have a substantial impact on Irving and the Mid-Cities, and this area was positioned to become, jointly, North Texas's third "Super City."[41]

When the Irving Chamber of Commerce traveled to the West Coast, council member Kenneth Reynolds was pleasantly surprised at how many people he met over there actually knew where Irving was. On previous trips to the West Coast, and other locations, he would invariably have to explain where Irving was. The publicity from Texas Stadium and the national television broadcasts of the Cowboys enabled people across the nation to locate Irving quickly.[42]

Texas Stadium was as important as any tool to the Irving Chamber of Commerce; in a way, it became their symbol. When representatives from the East Texas Chamber of Commerce visited Irving in 1971, they had their picture taken with Robert Power, with an image of Texas Stadium in the background. The Irving Chamber of Commerce also held dinners at Texas Stadium, inside the Inner Circle Club. What more impressive location could any organization hope to hold their dinners in? Prior to Texas Stadium, the chamber of commerce, along with other civic groups, had been relegated to holding its dinners in the drab Irving High School cafeteria.[43]

The increased activity of the chamber of commerce reflected the increasing deposits at Irving banks. In 1970, Irving laid claim to the two largest banks in Dallas County outside the city of Dallas—Irving Bank and Trust and Southwest Bank and Trust. That same year, the two banks reported "large increases in resources," a sign of the growth and apparent prosperity of Irving. Three years later, the total deposits in Irving banks added up to over $132 million, a $10 million increase over the previous year.[44]

Also in 1973, in the midst of the cold war, the Irving Chamber of Commerce hosted guests from the other side of the Iron Curtain. News of Texas Stadium reached across the globe. Members of the Moscow Dynamo soccer team were the guests of honor at a chamber of commerce board meeting, where the president of the Soviet Athletic Foundation presented the chamber with a team pennant. Perhaps the emerging status of Irving had softened feelings toward the Soviets, for only five years earlier there were plans to use part of Texas Stadium as a fallout shelter.[45]

Irving and Texas Stadium apparently had an effect on another sports team in another world capital. In 1971 the Washington Senators baseball team announced that it was moving to Arlington. When the Cowboys had decided to move to Irving, the eyes of the sports world became focused on Irving, not necessarily to watch a game but to see how this super-suburb would support a major athletic club. Evidently Irving had done at least as good a job as Dallas had. Though not directly causing the Senators to relocate, Irving demonstrated that a former bedroom community was capable of supporting a major franchise in a progressive manner. In addition, Irving helped to demonstrate the important role that surrounding cities could play in the metroplex. Major attractions and businesses located in cities surrounding Dallas and Fort Worth were able to contribute to the growth, profitability, and identity of the entire area without sacrificing their own individual identity.[46]

Tom Vandergriff had been attempting to locate a Major League Baseball team in Arlington for at least half a decade, and now this goal was being accomplished. Arlington was similar to Irving in many regards. Both were former bedroom communities that had evolved into vibrant communities, even suburban cities, within a booming metropolitan area. The road system around each of them had improved substantially, thanks to recent voter support for such bonds, and both Arlington and Irving stood to gain much from the new Dallas–Fort Worth Regional Airport. The Washington Senators were prepared to bolt an established city for a rising one. Irving demonstrated, with the Cowboys and Texas Stadium, that professional football could thrive in such a location. The Senators organization hoped for similar results in a similar setting, becoming the Texas Rangers, located in Arlington, in 1972.[47]

Irving had experienced growth prior to the construction of Texas Stadium and would continue to grow afterward. What Texas Stadium provided for Irving was a landmark, something that was recognizable not only citywide and countywide but nationwide and worldwide. In taking the Cowboys away from Dallas, Irving demonstrated that it was ready to be a major player and that it would readily compete with Dallas and Fort Worth. By arranging for the construction of Texas Stadium, Irving served as an example to other cities desirous of becoming important metropolitan centers.

NOTES

1. Bill Stevens, "Irvingites' Attitude Questioned," *Irving Daily News*, February 19, 1967; Sally Bell, "Opinions Differ on Cutting Tollroad." *Daily News Texan*, July 15, 1966. The *Daily News Texan* changed its name to the *Irving Daily News* in 1966.

2. Stevens, "Irvingites' Attitude Questioned."

3. James Ratteree, President, Irving Chamber of Commerce, letter to "George," September 19, 1966, Collection 6, Box 8, Irving Archives, Irving Public Library. Collection 6 is a compilation of clippings, brochures, and pictures collected by the Irving Chamber of Commerce through the years and donated to the Irving Archives. "Irving Delegation Attends State Fair." *Irving Daily News*, October 13, 1966.

4. "Land Prices Hurt City's Bid for Industrialization," *Daily News Texan*, July 3, 1966; "Irving at the Crossroads: A Community or a Bedroom?," *Irving Daily News*, February 5, 1967; "Canal Gets Boost," *Dallas Times Herald*, January 23, 1966; Jack Harkrider, "Trinity Canal, 'Face-Lifting Concern CofC," *Daily News Texan*, June 8, 1966.

5. Bill Stevens, "The Scratch Pad," *Daily News Texan*, April 7, 1966.

6. "Young Ideas Forum Termed a Success," *Irving Daily News*, October 27, 1966; "Airport Figures in Road Plans," *Irving Daily News*, April 26, 1967; "Regional Airport," undated newspaper clipping, Collection 6, Box 9, Irving Archives, Irving Public Library.

7. "Irving Scene," *Irving Daily News*, October 23, 1966.

8. "New Airport Explained," *Irving Daily News*, June 2, 1967; "Airport Riddle," *Irving Daily News*, June 4, 1967; "City Future up in the Air," *Irving Daily News* June 12, 1967.

9. *The Handbook of Texas Online*, "Cotton Bowl," www.tsha.utexas.edu/handbook /online/articles/view/CC/xxc1.html (accessed March 29, 2004). The Dallas Texans of the National Football League played only one year in Dallas before moving to Baltimore. The Dallas Texans of the American Football League were owned by Lamar Hunt and not affiliated with the previous Dallas Texans; this team would move to Kansas City after the 1962 season.

10. Don Smith, "Officials Say Crowd Too Big," *Dallas Morning News*, August 30, 1967; Gary Cartwright, "Kay Still Unshaken," *Dallas Morning News*, September 12, 1966; "Bond Payments Due on Nov. 1," *Dallas Morning News*, September 16, 1968.

11. Cartwright, "Kay Still Unshaken"; "Protection at Stadium to Continue," *Dallas Morning News*, September 12, 1967; Sam Blair, "Old Bowl in a Jam," *Dallas Morning News*, October 16, 1967; Gene Ormsby, "Ramsay 'Hopeful' Cowboys to Stay," *Dallas Morning News*, December 30, 1967; "Bond Payments Due on Nov. 1"; Sam Blair, "Open Season for Charm," *Dallas Morning News*, May 27, 1969.

12. Jane Wolfe, *The Murchisons* (New York: St. Martin's Press, 1989), 299–303; David Morgan, "Councilwoman Critical of Irving Stadium Deal," *Dallas Morning News*, December 25, 1967; Melvin Schuler, interviewer unknown, transcribed from tape recording, date unknown; Dan Matkin, interviewed by Yolanda Romero, transcribed from tape recording, January 15, 2003; and Robert Power, interviewed by Mrs. O. D. Bates, transcribed from tape recording, 1986, all from Oral History Collection, Irving Archives, Irving Public Library (hereafter IA-OHC). Although the interview with Schuler is undated, and the interviewer (probably himself) is unknown, the

interview does provide important details, most of which can be confirmed elsewhere. Still, the reader must be cautioned that Schuler's account, recorded over thirty years after the events, seems somewhat embellished by the years. Schuler was evidently a land speculator but, more important, he served on the planning and zoning commission with future Irving mayor Dan Matkin, according to Matkin's own interview.

13. Power, interview, IA-OHC; Wolfe, *Murchisons*, 299–303; Morgan, "Councilwoman Critical of Irving Stadium Deal."

14. For a detailed account of the construction of the Astrodome, see Travis Glenn Wise, "Space City and Shortstops: Baseball, Boosterism, and Houston's Coming of Age," Master's thesis, Stephen F. Austin State University, Nacogdoches, TX, 2010.

15. Mike Kingston, "Stadium Is Irving's—If," *Irving Daily News*, January 29, 1967; Bill Stevens, "Irvingites' Attitude Questioned," *Irving Daily News*, February 19, 1967.

16. Schuler interview, IA-OHC.

17. "Murchison Making a Drive," *Dallas Times Herald*, January 29, 1967; "Stadium Due Discussion by Council," *Irving Daily News*, undated, Collection 6, Box 9, Irving Archives, Irving Public Library.

18. Power, Bates, and Schuler interviews, IA-OHC.

19. Schuler interview, IA-OHC.

20. Joseph Rice, *Irving: A Texas Odyssey* (Chatsworth, CA: Windsor, 1989), 81–83; Robert Power, phone interview by the author, April 29, 2004.

21. Rice, *Irving*, 81–85.

22. Wolfe, *Murchisons*, 301–3; Gene Ormsby, "Ramsay 'Hopeful' Cowboys to Stay," *Dallas Morning News*, December 30, 1967.

23. "Bond Vote Set for Irving," *Dallas Times Herald*, January 13, 1968; Al Altwegg, "Problem on Stadium Bonds," *Dallas Morning News*, December 31, 1967.

24. Clint Murchison, "Thoughts and Otherwise," *Dallas Times Herald*, January 13, 1968; "Is Downtown Stadium in the Cards?," *Dallas Times Herald*, July 8, 1967.

25. "Irving OKs Stadium," *Washington Post*, January 17, 1969; Mike Kingston, "Kid Next Door Is Growing Up," *Dallas Morning News*, January 28, 1969; Power and Bates interviews, IA-OHC; "Power Accepts Hand-Off," *Irving Daily News*, January 21, 1969.

26. Bob St. John, "Texas Stadium: A Familiar Look," *Dallas Morning News*, January 26, 1969; Mike Kingston, "Texas Stadium: An Example for Chicago Mayor," *Irving Daily News*, January 11, 1968.

27. Kingston, "Kid Next Door Is Growing Up."

28. Irving Chamber of Commerce invitation, Collection 6, Box 10, Irving Archives, Irving Public Library; "Groundbreaking Event for Stadium," *Irving Daily News*, January 26, 1969.

29. "Braniff Tower UD Landmark," *Irving Daily News*, July 27, 1969; "Irvingites Rally for Acquisition of Texas Stadium," *Irving Daily News*, July 27, 1969.

30. John Elliott, "Stadium Construction Toured by Chamber Board Members," *Irving Daily News*, September 7, 1969; Mike Kingston, "Can You Get There—and Back—from Here?," *Dallas Morning News*, September 5, 1969; newspaper header, *Irving Daily News*, April 29, 1971.

31. Kara Rogge, "$6 Million in Bonds Approved for Stadium," *Irving Daily News*, November 21, 1969; Sam Blair, "Open Season for Charm," *Dallas Morning News*, May 27, 1969.

32. Gary Cartwright, "Clint Tosses Stadium Pass," *Dallas Morning News*, January 29, 1967; "Irving Hearing Set on 183 Freeway Plan," *Dallas Morning News*, February 14, 1969; Progress Report, Irving Chamber of Commerce, Collection 6, Box 11, Irving Archives, Irving Public Library; "Irving's 100,000th Family," *Irving Daily News*, August 19, 1969; "Irving's Gain," *Irving Daily News*, April 29, 1971.

33. Kara Rogge, "1969—Irving's Year of Growth," *Irving Daily News*, December 31, 1969.

34. "'Death Alley' Improved," *Irving Daily News*, April 21, 1967; "Council Drives Home Roads Plan," *Irving Daily News*, January 19, 1969.

35. Carl Freund, "Irving-Dallas Road Need Is Seen," *Dallas Morning News*, February 26, 1970; "Tollroad Would Touch Irving," *Irving Daily News*, November 14, 1971.

36. Mike Kingston, "Irving's Growing Pains Continue," *Dallas Morning News*, May 4, 1971; "Building Permits Update," *Dallas Times Herald*, April 6, 1969; "Chamber Members Return from Tour," *Irving Daily News*, October 13, 1969; "Murchison Boosts City," *Irving Daily News*, undated, Collection 6, Box 11, Irving Archives, Irving Public Library.

37. "Titche's, Sears Coming to Mall," *Irving Daily News*, March 19, 1969; "Carson Pirie Scott Arrives," *Irving Daily News*, June 27, 1969; "TIME-DC Inc. Erecting Facility," *Irving Daily News*, July 29, 1969.

38. "Storey Electric Relocates," *Irving Daily News*, November 6, 1969; John M. Findlay, *Magic Lands: Western Cityscapes and American Culture after 1940* (Berkeley: University of California Press, 1992), 52–105.

39. "Irving Expands Limits," *Dallas Times Herald*, April 28, 1974; 1987 City of Irving Annexation Map, Irving Archives, Irving Public Library; Rice, *Irving*, 84; "Planner, Annexation Are Young Ideas Forum Topic," *Irving Daily News*, January 24, 1967.

40. Patsi Aucoin, "Dallasite Asks Super C of C," *Irving Daily News*, December 18, 1969.

41. "C of C Invited to Dallas Meet," *Irving Daily News*, January 1, 1970; "Irving to Get 'Lion's Share,'" *Irving Daily News*, November 13, 1970.

42. "West Coast Tour Termed Success," *Irving Daily News*, October 10, 1972.

43. "Special Visitors," *Irving Daily News*, November 7, 1971; "Annual Chamber Dinner Slated Tomorrow Night," *Irving Daily News*, April 19, 1973; "Chamber Dinner Tomorrow Night," *Irving Daily News*, October 17, 1966.

44. "Thank You Irving," *Irving Daily News*, January 11, 1970; Larry L. Payton, "Deposits Climb $10 Million," *Irving Daily News*, April 4, 1973.

45. Progress Report, Irving Chamber of Commerce, Collection 6, Box 11, Irving Archives, Irving Public Library.

46. "Irvingites Excited over Senator Move," *Irving Daily News*, September 23, 1971.

47. Cartwright, "Clint Tosses Stadium Pass"; Larry Toth, "Road Bond Election Gets City's Support," *Irving Daily News*, May 9, 1975; "Irvingites Excited over Senator Move."

{ 7 }

"CAN I GET A YEE-HAW AND AN AMEN"

THE TEXAS COWBOY CHURCH MOVEMENT

JAKE McADAMS

This is the biggest religious movement in my lifetime. . . .
In this part of Texas, it's huge.

—KENNETH DENNY, Cowboy Church of Ellis County member

Since the 1970s, the United States, especially in the South and the West, has witnessed the creation and dramatic growth of the cowboy church movement. This movement has routinely made headlines as a unique brand of Christianity that encourages members to wear cowboy hats and ride horses to worship God. These popular media representations typically narrow on the perceived abnormalities of cowboy churches and primarily locate them in rural America. Nonmembers also tend to dismiss cowboy churches as the "Beer and Barbecue Church" where people play dress-up out in the country once a week. Nonetheless, the cowboy church movement remains a strong influence in contemporary society. Cowboy church members, called "cowboy Christians," are encouraged to incorporate images and practices of assumed American rural and western cultures into their worship services, in which they can exclaim both "Amen" and "Yee-haw" within the same sermon.[1]

Working with four Texas cowboy churches—the Baptist Cowboy Church of Ellis County in Waxahachie, the Impact Cowboy Church in Nacogdoches, Shepherd's Valley Cowboy Church in Cleburne, and the Cowboy Church of Henrietta—I compiled numerous oral histories to learn more about this movement and understand church members' attractions, motivations, and self-meaning making. The history and function of cowboy churches in Texas demonstrate that the cowboy church movement in Texas is, in fact, a suburban "seeker" church movement. Cowboy churches look to draw in a "seeker" by anchoring to a romanticized cowboy image and identity basis similar to that already discussed by Paul J. P. Sandul in this volume. Further, cowboy churches provide meaningful, sacred spaces—both materially and narratively—in which members, namely, (Texan) suburbanites, can partake in a seemingly authentic religious experience. These religious experiences fulfill seemingly personal sociocultural desires while helping to perform, propagate, and popularize them publicly to, and potentially for, others.

It is important to define a few terms related to the movement. "Evangelicalism" is the umbrella term that describes conservative Christianity and encompasses Protestant fundamentalists, charismatics, and those who reject those labels but self-identify as evangelical. It specifically refers here to Protestants who believe salvation is attained solely through an acknowledgment of personal sinfulness and atonement though Jesus, typically interpret the Bible literally, value converting nonmembers, and have a shared religious experience that includes similar modes of worship. Evangelicalism often, though not universally, engenders and demands that practitioners incorporate emotion and emotionalism into their religious practice and performance. As noted by scholars Donald Miller and Kimon Sargeant, this religious emotionalism is also prominent in the rise of evangelical seeker churches.[2] Two similar terms used in this chapter are "traditional evangelicalism" and "traditional churches." These terms refer to non–cowboy churches that affiliate with evangelical or mainline Protestant denominations including Baptists, Assemblies of God, Methodists, and Churches of Christ and do not primarily provide "contemporary" worship services.

"Charismatic" is another theological distinction made within the cowboy church movement. Occasionally referred to as "spirit-filled," charismatic Christians, or "charismatics," practice a highly emotional and physically expressive worship and believe in the personal reception and performance of "gifts of the Holy Spirit," namely, prophesying, healing, and speaking in tongues (uttering sounds that the speaker believes is a language spoken through him or her by God). It is also important to note that charismatics should be thought of as a subbranch of evangelicalism.

Finally, "unchurched" refers to people who currently do not attend a church. Many unchurched in the United States, especially those drawn to the cowboy church movement, have evangelical backgrounds but left their churches for personal or social reasons. Many unchurched are also considered religious "seekers," or those individuals not content with their current spiritual or religious worldview or system.

HISTORY

The contemporary cowboy church movement traces its origins to rodeo church services in the mid-1970s. By midcentury, rodeos and western culture were gaining popularity throughout the United States and Canada as communities and organizations recognized their abilities to function as sources of economic boosterism, displays of civic pride, and recreational opportunities. Despite rodeo's economic and cultural successes, many participants and attendees lived "un-Christian" lives. They consumed their Sundays with rodeo riding, not church going, participated in "ungodly" practices such as drinking alcohol, and did not meet churches' Sunday morning "dress codes." With traditional churches proving irrelevant and incompatible with their lives, such rodeo participants remained unchurched.[3]

Within this seemingly godless environment, Christian rodeo participants began holding church services at rodeos. These early, nondenominational services provided an informal setting at the rodeo arena for Christian rodeo participants to worship on Sundays before that night's competition. Many prominent rodeo athletes, including Wilbur Plaugher and Mark Schriker, held services at the rodeos in which they participated. These services quickly gave rise to three Christian rodeo cowboy associations in the 1970s: the Baptist "Cowboys for Christ" in the eastern United States; and the charismatic "Rodeo Cowboys Ministry" and "Fellowship of Christian Cowboys" along the western rodeo circuits. These early "come-as-you-are" rodeo services attracted Christian and non-Christian rodeo participants and spectators and saw as many as three hundred attendees at a single service.[4]

In the mid-1980s, rodeo church services became more prevalent as college rodeo athletes entered the professional circuit. Many of these young professional rodeo riders participated in the evangelical revival of the 1970s, which stressed the literal interpretation of the Bible and importance of the Holy Spirit and sought to provide attractive, culturally relevant liturgies and teachings. These youth, such as Ron Weaver, Paul Scholtz, and Jeff Copenhaver, transformed the informal rodeo services of the 1970s into a more mainstream practice and gained ministerial commissions by the Assemblies of God and other denominations. Rodeo services grew significant enough that Weaver and Copenhaver agreed in 1985 to hold a service for attendees of the National Finals Rodeo in Las Vegas. Billy Bob Barnett, owner of Billy Bob's Texas honky-tonk in Fort Worth, attended this service. Barnett appreciated Weaver and Copenhaver's message so much that he invited Copenhaver to hold services at the indoor bull-riding ring at Billy Bob's during the 1986 Fort Worth Stock Show and Rodeo. More than two hundred attended these services, and Barnett signed a two-year contract with Copenhaver to hold weekly church services in the honky-tonk, thus creating the first stationary contemporary cowboy church in the world.[5]

Despite marginal successes outside Texas, most early cowboy churches were located in the Lone Star State. Larry Miller's Cowboy Church of Henrietta, located southeast of Wichita Falls, and Russ Weaver's Shepherd's Valley Cowboy Church in Alvarado, about fifteen miles east of Cleburne, south of Fort Worth, stand out among these early cowboy churches. Miller began his church at Henrietta in the local horse sale barn in 1992. After a few years of wavering attendance, Miller obtained an ordination from Copenhaver, thus legitimizing his nondenominational ministry, and the group put up a new building on a hill overlooking U.S. Highway 287 outside Henrietta. The group wanted to advertise their new location, so they created the Cowboy and Cross icon to display their Christian belief and cultural identity. After this move, the Cowboy Church of Henrietta quickly grew to 550 members by 2000.[6]

Assemblies of God pastor Russ Weaver and Shepherd's Valley followed a different path. After almost two decades of rodeo ministry, Weaver sought to settle down in 1996. After Weaver failed to find a place to hold cowboy church services, Baptist radio personality Dawson McAlister instigated a partnership to rent the Walt Garrison Arena in Benbrook, Texas, for a token amount and allow Weaver to pastor the church. The church maintained sporadic attendance for several years until Weaver relocated to a new building outside Cleburne. At the new location, Shepherd's Valley grew much larger and began to plant other local cowboy churches. These early developments laid the foundation for the cowboy church movement, but the movement maintained mixed success until the twenty-first century.[7]

In 2000, Baptist church planter Ron Nolen created the first Baptist-affiliated cowboy church in Waxahachie, south of Dallas, and the movement rapidly spread. After attending rodeos with his son, Nolen recognized the lack of church participation among rodeo participants. Aware of previous cowboy churches as well as contemporary seeker-sensitive tactics to reach the unchurched, Nolen sought to institute a church model that attracted rodeo participants and other self-affiliates of what he called the "Western Heritage Culture." Intending to create a more inclusive church community, Nolen identified his audience broadly, but the Western Heritage Culture generally refers to (white) political conservatives who exhibit "rural," "western," or "country" cultural identifiers, including participation in rodeos, watching western movies, dressing in cowboy hats and boots, and listening to country music. In other words, Nolen wanted to attract anyone comfortable with John Wayne and uncomfortable with traditional churches. Working with the Baptist General Convention of Texas and the First Baptist Church of Waxahachie, Nolen gathered financial support and held the first Cowboy Church of Ellis County (CCEC) service in the Ellis County Expo Building on March 5, 2000. Approximately three hundred attended this service, and Nolen continued to hold weekly services in the expansive Expo building for the next year and a half, drawing people from within a sixty-mile radius of Waxahachie.[8]

Although affiliated with the Baptist General Convention of Texas, Nolen broke from traditional Baptist organizational practices by appointing elders and lay pastors instead of deacons. Following earlier cowboy churches' models, Nolen also refused to take a structured offering during services. Maintaining a steady three hundred–person attendance, Nolen and CCEC leadership decided to construct their own church building and rodeo arena approximately two hundred yards west of the Expo center in 2001. Having firmly established the CCEC and become enthused about this new type of church, Nolen left the CCEC and founded the Texas Fellowship of Cowboy Churches, enabling him to plant Baptist cowboy churches elsewhere in the state.[9]

CCEC leadership hired Gary Morgan to fill Nolen's position. Under Morgan, the CCEC continued to work with Nolen and fine-tuned a "Low Barriers Model" (see below) church structure of the CCEC. Morgan explains that this model removes the formalities often associated with traditional evangelical "religion" (such as structured contributions, professional dress, and supposed feminized worship) and cultivates an environment attractive to unchurched individuals, specifically men, who self-identify with the Western Heritage Culture.[10]

After 2000 the cowboy church movement experienced dramatic growth throughout Texas, the United States, and the world. Acknowledging popular interest in the movement, Nolen and his renamed American Fellowship of Cowboy Churches hosted the first cowboy church training session in 2004 to educate interested people about the Low Barriers Model, as well as how to establish a cowboy church. These training sessions proved immensely effective, and many Baptists founded new cowboy churches and many Texans found new "church homes." Satellite television station RFD-TV contracted Russ Weaver and Suzie McEntire to host a weekly, thirty-minute televangelism program titled *Cowboy Church TV* in 2007. This program won national attention for the cowboy church movement and increased cowboy church attendance. The online cowboy church movement directory, CowboyChurch.net, listed more than 850 cowboy ministries in the United States in 2013, which include stationary cowboy churches, rodeo ministries, and music ministries. Furthermore, the cowboy church movement claims churches in Canada, Australia, the Philippines, Russia, and Kenya, and countless people stream cowboy church sermons online. Despite its international expanses, the cowboy church movement remains strongest in Texas, which claims at least 160 cowboy churches, including two Spanish-speaking congregations.[11]

COWBOY CHRISTIANITY

Cowboy churches provide a unique worship experience for all attendees, complete with country-style music, come-as-you-are dress, and, commonly, refreshments. Although Sunday gatherings center around a twenty- to thirty-minute message similar to traditional churches, many cowboy Christians explain that the worship at cowboy churches seems more "genuine," "authentic," and "real" compared to churches they previously attended. Briefly reviewing the cowboy church movement's theology and liturgy helps determine where the movement fits culturally and how members integrate personal, social, and geographical values into their worldviews. Although these observations do not explain every individual cowboy church or cowboy Christian, they are common in many Texas cowboy churches.

Explaining cowboy Christian theology is deceptively simple: each cowboy church preaches a doctrine that aligns with its denominational affiliation. Since the cowboy church movement is a transdenominational movement and not a single, unified denomination, individual cowboy churches commonly affiliate with larger, established Christian denominations. In Texas, the Baptist General Convention of Texas claims the most cowboy churches, then the Southern Baptist Convention, then charismatic denominations, including the Assemblies of God. There are also several nondenominational churches, which are typically charismatic leaning. These denominations' theologies, as well as various cowboy church statements of faith, proclaim that there is one God, human salvation is attained only through belief in Jesus Christ, the Holy Spirit is integral to Christians' lives, the Bible is the inspired word of God and should be interpreted literally, Christians should be baptized to fulfill Jesus's command and demonstrate their salvation, there is a literal Heaven and Hell, and there will be a physical resurrection at a second coming of Jesus.[12] Additionally, many cowboy church pastors have at least some seminary training.

The difference between cowboy churches and traditional evangelical churches is the former's congregational organization methodology and liturgical framework—the aforementioned Low Barriers Model. Cowboy Christians explain that many unchurched equate religion with the cause of wars and the motivation to fill coffers instead of helping individuals. Others contend that traditional churches' expectations of church dress and putting on a "big plastic banana smile" produce a religion that lacks heart-felt commitment to Jesus Christ and fellow church members. The Low Barriers Model attempts to remove these perceived barriers of traditional American religion, namely, pretentiousness, which previously turned many cowboy Christians away from Christianity.[13]

Morgan explains that the model is based on what he calls the "essence" of the cowboy church movement—not expecting people to act like Christians before they are saved. This means that cowboy churches, especially church leaders, do not expect people to conform to commonly held standards of "Christian living," which include sobriety, modest dress, and unadulterated monogamy, before "accepting Jesus as their savior." Cowboy Christians explain that this theological premise enables cowboy churches to reach the unchurched, whom traditional churches have failed to reach because they expect people to conform to their moral and social standards *before* converting to Christianity.[14]

The Low Barriers Model attempts to frame Christianity as an intimate relationship with Jesus Christ that does not separate members' lives outside of church attendance from their spiritual journeys. This means cowboy churches encourage people to integrate Jesus and God into their daily lives but also to challenge traditional notions of sacred and profane. A striking example is cowboy Christians' dress during worship services. Commonly, members wear blue jeans, western style shirts, and a cowboy hat during Sunday gatherings. Many outside the movement think pastors and members wearing hats during services is symptomatic of irreverence bordering on blasphemy. Cowboy Christians, however, respond that men remove their hats during prayer to show their reverence toward God but wear hats during services because it is comfortable for attendees who identify with the Western Heritage Culture. Additionally, pastors use vernacular phrases during sermons to explain biblical interpretations instead of seminary jargon foreign to lay members. Furthermore, many cowboy churches meet in metal, barnlike buildings, thus removing the initial visual barriers many unchurched experience in traditional houses of worship as well as providing a culturally comfortable, practical place to gather.[15]

In this context, the cowboy church movement can be understood as a "vulgar" religion. This does not imply cowboy Christianity's sinfulness in any way, but rather cowboy Christianity's comfortable use, even sacralization, of the profane and signifies it as a religion belonging to "common people." The Low Barriers Model does not appeal to highbrow, traditionalistic American Christians but rather seeks to meet contemporary Americans' sociocultural tastes, offer relevant guidance for their lives, and provide individuals a friend, confidant, and savior in Jesus Christ, which to cowboy Christians represents a more authentic Christianity.

SUBURBAN COWBOY

An observant reader has probably noticed several common suburban themes appearing within the cowboy church movement already, including its synthesized "western" lifestyle. To explore the cowboy church movement's suburban connections further, it is important to define suburbia and examine how it is manifested in contemporary America.

Physical geographers aside, many define suburbia through cultural characteristics. Suburban historian Kenneth Jackson, for example, pronounces that suburbia is a "state of mind based on imagery and symbolism." Suburban historian Paul J. P. Sandul further explains that reducing suburbia to a physical location "risks reducing the importance of imagery and imagination, of storied space. Suburbia is as much a cultural symbol and

intellectual creation, even lived space, as a geographical, material place."
Moreover, scholars indicate that suburbia, not the rural America that suburbia appropriated, has become the "American dream" where potential
riches are made and traditional values are both realized and defended.
Suburbanism, as many Texans encounter it, is the combination of creative
imagination and land development policies that portray residents' values
of hard work, individualism, purity, and property ownership as definitions
of proper citizenship. These values allow them to remain connected to the
conservative middle class and its so-called traditional values.[16]

Three key factors define the cowboy church movement as suburban.
For one, the location of meetinghouses and cowboy Christians' residences
illustrates that suburbs, not rural areas, are the most influential places in
the movement. Additionally, characterizing the movement as a seeker
church philosophically connects it to larger suburban trends. Further,
cowboy Christians' definitions of suburbia and rural sacralization indicate
that the movement rests upon an ideal of rurality, which is intimately
linked to suburbia. These factors routinely manifest in cowboy Christians'
built environments, oral histories, and religious messages, thus helping
define the cowboy church movement as an American suburban product
and project.

Geographically locating the cowboy church movement presents an
obvious, yet often misinterpreted, indication of its suburban connections.
Though the rodeo services are difficult to define as suburban because
of their transience, the cowboy church movement beginning with Jeff
Copenhaver's church at Billy Bob's honky-tonk physically locates cowboy
churches within suburbia. As the movement grew, many early churches
began in suburban areas such as Russ Weaver's Shepherd's Valley Cowboy
Church. Currently, the most prominent Texas cowboy churches, including the Cowboy Church of Ellis County, Lone Star Cowboy Church in
Montgomery (north of Houston), and the Arena of Life Cowboy Church
in Amarillo, are all located in suburbs. Additionally, many of the cowboy
church fellowships and training programs are headquartered in suburban
locales, including the American Fellowship of Cowboy Churches and the
Cowboy Church Fellowship of the Assemblies of God. In other words,
despite emitting rural imagery, the movement's leadership is predominately located in suburbs. Explanations of this suburban growth receive
more attention below, but the simple fact of the movement's prevalence
within suburban areas indicates its suburban bona fides.[17]

Although many meetinghouses are located in suburbs, other observers note cowboy churches in rural areas. The obvious critique is that these meetinghouses are actually located in exurbs, the developed areas on the rural fringes. The often more accurate critique, though, is that although a meetinghouse is rurally located, a large percentage of its members are suburbanites. The vast majority of cowboy Christians interviewed live either inside city limits or, more often, in developed neighborhoods outside the city limits. Additionally, almost all interviewed daily drove into town to work and shop instead of working around their house—the classic suburban characteristic. This evidence indicates that, even if the meetinghouse is rurally located, the church is not necessarily rural in terms of membership.[18]

Another key factor indicating the cowboy church movement's suburban characteristics is categorizing it as a seeker church movement. Historians Dolores Hayden and Robert Fishman indicate that suburbs historically represented asylum for evangelicals—a middle landscape in which to live full lives, complete with urban conveniences without interacting with the perceived corruption and sin of cities. Cowboy Christians express a similar urban phobia, stating that the farther removed from urban areas, the "simpler" and, ultimately, more pure one's life becomes. To cowboy Christians, suburban areas, which they commonly refer to as rural, resemble a place of near-perfection, almost like the Garden of Eden.[19]

Religious scholars including Donald Miller, Mark Shibley, and Gretchen Buggeln also note that churches, especially seeker churches, are key features within American suburban landscapes. First appearing in the 1960s as suburban youth became disenchanted with traditional churches' seemingly outdated theologies and liturgies, seeker churches provide socioculturally relevant worship to attract unchurched who self-affiliate with specific cultures. Suburbs were the ideal place for such churches because they had ever-increasing populations, were sufficiently distanced from "sinful" cities, and had existing infrastructure that enabled leaders to promote their churches. Moreover, suburbs felt comfortable and seeker church leaders wanted members to feel comfortable at church. Recognizing this, seeker churches adopted suburban architecture, and suburbs seemed beset by shopping mall–like churches with recreational childcare facilities and spaces for church youth groups designed to make their suburban memberships, particularly families, feel more comfortable.[20] Additionally, the rise of megachurches—such as the Cowboy Church of Ellis

County, the Lone Star Cowboy Church in Montgomery, and the Arena of Life Cowboy Church in Amarillo—was possible only because of the population concentration in suburban areas. The seeker church movement undoubtedly, and possibly the entirety of American Christianity, became a suburban performance in the late twentieth century.[21]

Although some seeker churches wholly embrace American suburbia, the theologically conservative seeker churches, which the cowboy church movement resembles, started as a negative critique of suburban America. Miller explains that the suburban founders of conservative seeker churches saw modernity, as championed by the suburb, as banal, insincere, even sinful and sought a more "authentic" Christianity. By championing the nuclear family, emphasizing the individual, and creating strong communities of believers, these conservative seeker churches provided a stark alternative to the perceived sterility and lack of community in contemporary, commercialized, faceless America. These churches also sacralized the vernacular through their messages, dress, and liturgies, which aligns with contemporary suburbanites increasingly adopting and glorifying a working-class, blue-collar work ethic and romanticizing distinctly rural themes. Nonetheless, these churches began and grew in suburbs by attracting like-minded suburbanites and providing both nostalgia through their conservative theologies and an embrace of a modern, even futuristic, culture through their relaxed dress and cutting-edge music.[22]

The cowboy church movement follows this model. Not only do cowboy churches begin in and largely attract suburbanites, they also criticize U.S. suburbanism and embrace manifestly rural images, hence the cowboy and his horse. Cowboy Christians indicate disenchantment with traditional churches and construct spaces to make people comfortable. Certainly, cowboy churches sacralize the vernacular and profane and denounce other religious options as superficial and suburbia as pretentious. The cowboy church movement's cultivation of strong, active communities replicates other suburban critiques expressed by conservative seeker churches and new suburban communities such as the New Urbanists. Recognizing cowboy churches as a seeker church movement indicates that they are a religious manifestation of contemporary land development trends to expand the reach of suburbs and enable suburbanites to make sense of their lives. The cowboy church movement, then, presents a religion that enables suburbanites seemingly to withdraw further from developed areas by evoking imagery of the cowboy and cultivating a sense of the rural.[23]

This cultivated mentality is a third indicator that the cowboy church movement is suburban. In fact, it is the movement's linchpin, combining notions of Christian, cowboy, and American identity with cowboy Christians' uncomfortable suburban realities. Through shared values and performances, the cowboy church movement provides a strong community of memory that shapes cowboy Christians' identities around notions of rurality.

In *Making the San Fernando Valley*, urban geographer Laura Barraclough examines the horse suburbs of Southern California to describe the process of "rural urbanism" as the "production of rural landscapes by the urban state, capital, and other urban interests" to create and negotiate "American identity" through expressions of "rurality." According to Barraclough, rurality is a mindset constructed by combining rural exceptionalism with heroic white myths, such as the cowboy. Rural urban developers, then, construct spaces that embody the middle class's politically conservative and evangelical values, specifically large-lot homesteads that aesthetically evoke a rural atmosphere. This cultural landscape exemplifies a sense of privacy, property ownership, naturalized purity, and individualism, which is exclusive and mirrors other suburban trends such as gated communities. Contradictory to their political conservative ideals of rugged individualism and small government, residents often express a sense of entitlement because they are preserving supposed genuine "American heritage." Barraclough further defines rurality as the belief "that rural shapes human character, community dynamics, and the social order in superior ways, instilling in children and adults strong moral character, independence, humility, concern for community, and commitment to democratic participation. By contrast, [rural suburbanites] believe that urban and suburban lifestyles lead to moral weakness, criminality, greed, corruption, poor decision-making abilities, political apathy, and a lack of concern for one's neighbors, animals and the environment." Texan cowboy Christians' identities mirror these Californians, which, despite any groans otherwise, places the movement squarely within a suburban context.[24]

The cowboy church movement capitalizes on concepts of rurality, as illustrated by the built landscape and cowboy Christians' espoused values. The common cowboy church meetinghouse, constructed on a large lot and architecturally resembling buildings seen in rural areas, creates a rural atmosphere and provides a sense of privacy. Cowboy Christians also decorate the buildings' interiors to replicate the countryside and create a seemingly rural atmosphere in which to worship God. Furthermore, cowboy churches' rodeo arenas provide members safe locations to perform assumed rural performances with the comforts of more "civilized" areas, such as kitchens and air-conditioned bathrooms.

Away from the meetinghouse, many cowboy Christians living in developed areas outside city limits own ranchettes (large-lot residences) within developed neighborhoods. Similar to the meetinghouses, these ranchettes provide the semblance of rural living, complete with the cherished value of privacy, while maintaining suburban qualities such as municipal infrastructure and easy commutes to work. These qualities, plus the token reality of cowboy Christians who actually do live in rural areas, encourage cowboy Christians to identify as rural while remaining largely suburban.[25]

Cowboy Christians further identify as rural through attributed values. When asked to differentiate between urban, suburban, and rural areas, most cowboy Christians define these areas as a set of values or qualities, not geographic characteristics. Virtues such as "simple," "slow-paced," "peaceful," and "straightforward" top the rural list. Interviewees generally agree that raising children in rural areas away from cities and suburbs is ideal because they interact with nature and participate in animal husbandry. This supposedly teaches children the virtue of hard work and how to cope with loss and, ultimately, makes them better adults.[26]

Cowboy Christians sufficiently romanticize the rural and rural residents, and cowboy pastors routinely provide biblical "proof" of these concepts. A sermon at a cowboy church often depicts God, specifically Jesus, as a rural resident and shepherd. Using the Psalms and passages of Jesus as a shepherd, pastors construct guiding narratives that illustrate God endorsing the idealized rural life. The Bible also proves to cowboy Christians that other supposed rural values like hard work, individualism, and traditional gender roles are legitimate. Finally, cowboy Christians blame city dwellers and other suburbanites for destroying the nation's morality and their beloved rural landscape as sprawling development continually encroaches upon the rural.[27]

The cowboy church movement represents to cowboy Christians the last frontier, the last piece of rural America uncorrupted by contemporary society. After their failed physical attempts to claim, or reclaim, an authentic rurality through politics or land development, many suburbanites depend upon the cowboy church movement to provide a community of memory that cherishes the evaporated rural life. Therefore, although it provides a genuine worship experience that members receive great benefit from, the cowboy church movement promotes an idealized and mythic way of life.

CONCLUSION

The cowboy church movement operates on an international scale and therefore demands ever more consideration, scrutiny, and analysis. Not even a dream fifty years ago, cowboy churches currently contend for space in the religious marketplace and sprawling suburbia. They evoke notions of rurality and vow to meet people on their levels, which attract many to a life of Christian service. Although cowboy Christians construct their cultural appeals around national myths, they provide members a genuine religious and social experience that necessitates acknowledgment.

The cowboy church movement is primarily a religious movement in that it emphasizes, above all else, that individuals gain salvation in Jesus Christ through baptism and must live a Christian life. Cowboy Christians attempt to correct the perceived failures of traditional churches through their Low Barriers Model and socioculturally relevant liturgy and applicable theology and have met measured success at reaching the unchurched.

The movement, though, is also thoroughly suburban. This does not disregard the cowboy churches located in rural areas or the cowboy Christians who live in rural areas and have agricultural jobs. It is impossible, however, to overlook that the movement began and its leadership and members largely remain within suburbia, specifically the Dallas–Fort Worth Metroplex. From these suburban headquarters, cowboy Christians assist established cowboy churches and plant new groups throughout the world and can influence anyone with internet access. Additionally, the simple fact that the cowboy church movement is a seeker church movement directly ties it into national suburban cultural and religious trends. Furthermore, the glorification of the rural and construction of notions of rurality seemingly pose cowboy churches as a last frontier where suburbanites can escape the sins of the city and retreat with God to the country. Put differently, even if not physically rooted in suburbia, the cowboy church movement, as members claim it stands for, appropriates tropes and ideals that have been at the heart of suburban rhetoric and growth in the United States since the nineteenth century.

Before I send readers down the proverbial dusty trail, country music legend Tanya Tucker's 1978 "Texas (When I Die)" comes to mind. Tucker brings together a surprising number of factors at play within the cowboy church movement as she sings, "When I die I may not go to heaven / I don't know if they let cowboys in / If they don't, just let me go to Texas, boy! / Texas is as close as I've been." This song reflects both the skepticism and idealization that exists within cowboy churches. Tucker indicates the struggle between the sins of the "cowboy life" and Christian salvation and ponders the exact question posed by many contemporary cowboy Christians: Will I, as a cowboy, go to heaven? Additionally, Tucker's description of Texas as heaven on earth is startlingly similar to cowboy Christians' descriptions of heaven as full of pastures and God riding the range. Unlike Tucker, though, cowboy Christians have assurances in their "real" Christianity that God does let cowboys in, and that Peter will greet them at the pearly gates with a big "yee-haw" and an "amen."

NOTES

This chapter is based in part on interviews between the author and eighteen cowboy Christians. Though not comprehensive, this interview sampling provides a demographically and geographically diverse perspective on the movement in Texas. All collected oral histories and observation data are publicly available at "Texas Cowboy Church Oral History Project," East Texas Research Center, Stephen F. Austin State University, Nacogdoches, TX (hereafter TCCP), http://digital.sfasu.edu/cdm/search/collection /OH/searchterm/ Texas%20Cowboy%20Church%20oral%20History%20Project/field /collea/mode/all/conn/and/order/nosort. The chapter epigraph is from Kenneth and Cynthia Denny, interviewed by Jake McAdams, Waxahachie, TX, July 9, 2013.

1. Linda Owen, "Worship at the O. K. Corral: Cowboy Churches Shape Their Ministries for the Western at Heart," *Christianity Today*, September 2003, 62–63; John Burnett, "Cowboy Church: With Rodeo Arena, They 'Do Church Different,'" National Public Radio, last modified September 1, 2013, www.npr.org/2013/09/01/ 217268202 /cowboy-church-with-rodeo-arena-they-do-church-different; Gary Morgan, interviewed by Jake McAdams, Waxahachie, TX, July 11, 2013, TCCP.

2. Donald Miller, *Reinventing American Protestantism: Christianity in the New Millennium* (Berkeley: University of California Press, 1999), 85–90; Kimon Sargeant, *Seeker Churches: Promoting Traditional Religion in a Nontraditional Way* (New Brunswick, NJ: Rutgers University Press, 2000), 40, 64, 172.

3. Michael Allen, *Rodeo Cowboys in the North American Imagination* (Reno: University of Nevada Press, 1998), 25, 35; Russ Weaver, interviewed by Jake McAdams, Egan, TX, July 12, 2013, TCCP.

4. "Tribute to Ted," Cowboys for Christ, https://www.cowboysforchrist.org /tribute.php (accessed August 19, 2013); "Cowboy Church to Feature Hall-of-Famer Wilbur Plaugher," *Sierra Star*, last modified May 1, 2013, www.sierrastar.com /2013/05/01/62395/cowboy-church-to-feature-hall.html; Weaver interview, TCCP.

5. Weaver interview, TCCP; Jeff Copenhaver, "America's First Cowboy Church," *Copenhaver Ministries*, http://jeffcopenhaver.com/firstcowboychurch.html (accessed August 22, 2013).

6. Larry Miller, interviewed by Jake McAdams, Porter, TX, September 19, 2013, TCCP.

7. Weaver interview, TCCP.

8. Morgan interview, TCCP; Gregg Horn, "The History of the TFCC/AFCC" (unpublished, American Fellowship of Cowboy Churches, Waxahachie, TX), 1.

9. Morgan interview, TCCP.

10. Morgan interview, TCCP.

11. Horn, "History of the TFCC/AFCC," 2; "Hosts," *Cowboy Church-TV*, http:// cowboychurch.tv/hosts (accessed July 29, 2013); "Founders," *Cowboy Ministers Network*, http://cowboyministersnetwork.org/founders.htm (accessed August 28, 2013); "Directories," *CowboyChurch.net*, www.cowboychurch.net/directories.html (accessed August 28, 2013); "Ministry Directory," *International Cowboy Church Alliance Network*, http://iccanlink.ning.com/page/church-directory (accessed August 28, 2013); "AOL Churches," *Arena of Life*, www.amarillocowboychurch.org/aolchurches .html (accessed August 28, 2013); "Missions Work," *Jeff Copenhaver Ministries*, http:// jeffcopenhaver.com/missionswork.html (accessed August 28, 2013).

12. Based, for example, on statements of faith from the Cowboy Church of Ellis County, Impact Cowboy Church of Nacogdoches, and Shepherd's Valley Cowboy Church. http://www.shepherdsvalley.com/egan/index.php/about/belief.

13. Although Gary Morgan and Ron Nolen coined the term "Low Barriers Model," it refers here to the general tenets of the shared methodology within the movement and not solely the American Fellowship of Cowboy Churches' official model. Stan King, interviewed by Jake McAdams, Nacogdoches, TX, June 4, 2013, TCCP; Morgan interview, TCCP.

14. Morgan, Weaver, Stan King interviews, TCCP.

15. Tommy and Vivian Sublett, interviewed by Jake McAdams, Nacogdoches, TX, June 17, 2013, TCCP; Rodney and Kristi Spencer, interviewed by Jake McAdams, Nacogdoches, TX, July 2, 2013, TCCP; Stan King interview, TCCP.

16. Kenneth T. Jackson, *Crabgrass Frontier: The Suburbanization of the United States* (New York: Oxford University Press, 1985), 5, 272, 288; Paul J. P. Sandul, *California Dreaming: Boosterism, Memory, and Rural Suburbs in the Golden State* (Morgantown: West Virginia University Press, 2014), 86; Dolores Hayden, *Building Suburbia: Green Fields and Urban Growth, 1820–2000* (New York: Vintage Books, 2004), 38–41.

17. "Texas Directory," *CowboyChurch.net*, www.cowboychurch.net/texas.html (accessed August 28, 2013).

18. Sarah Moczygemba, "Rounding Up Christian Cowboys: Myth, Masculinity, and Identity in Two Texas Congregations," Master's thesis, University of Florida, 2013, 35, 46.

19. Robert Fishman, *Bourgeois Utopias: The Rise and Fall of Suburbia* (New York: Basic Books, 1987), 53; Hayden, *Building Suburbia*, 6; Caye King, interviewed by Jake McAdams, Nacogdoches, TX, June 17, 2013; Rex Crenshaw, interviewed by Jake McAdams, Nacogdoches, TX, June 9, 2013; Spencer interview; Robert Wilson, interviewed by Jake McAdams, Waxahachie, TX, July 29, 2013, all interviews in TCCP.

20. Gretchen Buggeln, "Spaces for Youth in Suburban Protestant Churches," in John Archer, Paul J. P. Sandul, and Katherine Solomonson, eds., *Making Suburbia: New Histories of Everyday America* (Minneapolis: University of Minnesota Press, 2015), 227–39.

21. Miller, *Reinventing American Protestantism*, 4, 17; Mark A. Shibley, *Resurgent Evangelicalism in the United States: Mapping Cultural Change since 1970* (Columbia: University of South Carolina Press, 1996), 5.

22. Miller, *Reinventing American Protestantism*, 1, 20, 110.

23. Weaver, Caye King, Spencer, Denny, and Wilson interviews, TCCP; John W. Williford Jr., "Ethereal Cowboy Way: An Ethnographic Study of Cowboy Churches Today," Ph.D. dissertation, Regent University, 2011, 30–31.

24. Laura A. Barraclough, *Making the San Fernando Valley: Rural Landscape, Urban Development, and White Privilege* (Athens: University of Georgia Press, 2011), 2 (quote), 8–9, 138–41, 206 (quote).

25. Morgan interview, TCCP.

26. Stan King, Caye King, Crenshaw, Spencer, Morgan, Wilson, Weaver interviews, TCCP.

27. Impact Cowboy Church, worship service, November 28, 2012; Impact Cowboy Church, worship service, May 29, 2013; Cowboy Church of Ellis County, worship service, Waxahachie, TX, July 8, 2013; Stan King, Sublett, Morgan interviews, TCCP.

{ 8 }

KEEPING THE CITY SUBURBAN

DENSITY AND THE PARADOX OF ENVIRONMENTAL
PROTECTIONISM IN AUSTIN

ANDREW BUSCH

It may come as little surprise that among Texas's cities Austin is best known for its environmental qualities and, perhaps more so, for its environmental spirit. Over the past three decades the capital city of Texas has become a national and global leader in numerous environmental categories. Publicly owned utility Austin Energy leads the nation in renewable energy sales. It also offers a robust rebate program for environmentally friendly upgrades. The city council anticipates that Austin will be carbon neutral by 2020. It has also been recently credited as a leader in clean technology, park and trail space, and sustainable urban planning. Austin has managed to achieve these accolades despite consistently strong demographic growth; even in fast-growing Texas, Austin continues to add a higher percentage of residents than other large Texas cities, and its economic growth continues to far outpace even booming Dallas and Houston.[1]

Much of the city of Austin's environmental stewardship is due to the agency of people dedicated to maintaining Austin's unique environmental and social qualities. In the 1970s and 1980s citizens organized grassroots environmental campaigns, neighborhood organizations, and public planning initiatives and battled developers in courts and on zoning boards. By the late 1980s citizens from around the city had coalesced in defense of Barton Springs, a popular leisure site that was under threat from development-related pollution. The springs became a symbol of the city's unique natural qualities but also of its unique social consciousness. "The pangs of loss are not, as in Dallas and Houston," one writer opined in 1990, "for the days when the skies were full of construction cranes . . . but for things that are irretrievably gone: the view west from Mount Bonnell of pristine hills, now carved into subdivisions; the low water crossing on a country lane, now turned into an outer loop; the steep hill on the edge of town, now bulldozed into a stubby mesa."[2]

In the 1990s their efforts led to a massive overhaul of the city's development philosophy. In the early 1990s environmentalists convinced the city council to pass a number of strong developmental restraints in sensitive areas. By the late 1990s they had elected the "Green Council," a full slate of city council members who valued the environment and promised to defend it. Mayor Kirk Watson instituted the Smart Growth Initiative, based on progressive urban planning ideology, which incentivized development in the central urban core by offering subsidies and creating widespread zoning changes in central Austin. Two large bond packages paid for developer subsidies and infrastructural improvements in central Austin and paid for land rights to ensure that more than 30,000 acres of pristine forest over the Edwards Aquifer on the city's western edge would remain undeveloped.

Yet these victories, and the arguments that supported them, contained two paradoxes that had wide-ranging implications for Austin's demographic and physical landscape over the coming years. First, though residents attacked suburban sprawl—new office parks on the Edwards Aquifer, the MOPAC highway that carved through west Austin, large residential subdivisions throughout northwest and southwest Austin, and the developers and city officials who profited from growth (at the expense of the city's natural environment and sense of place)—they were also critical of potential zoning changes that would increase density in their neighborhoods. Many older neighborhoods were still protected by restrictive covenants that likewise limited density. Thus, neighborhood groups found themselves in a precarious rhetorical position where they either had to encourage development in other areas or denounce growth entirely. Most chose the second option, where logic dictated that the "growth machine"— large-scale developers and the politicians who supported them—was the primary culprit in the city's precipitous decline. Yet the city continued to grow rapidly because of a booming high-tech industry and a robust cultural sector. If growth continued, sprawl was curtailed, and density did not increase, where would new residents live?

The second paradox, which was less overt, centered on the relationships among class, race, and environment. Sociologist Scott Swearingen, a participant observer who chronicled the anti-development movement in Austin in his book *Environmental City*, captures the paradox while not addressing the tension embedded in it. For Swearingen, Austin's environmentalists were largely drawn from the white middle-class, educated professionals working in government or at the university who did not benefit from increased development or from economic growth. To them, nature was valuable because it tied them to place but also because it afforded many recreational opportunities. Yet environmentalists have long been accused of treating nature as sacred, often at the expense of underprivileged communities.[3] In fighting against new housing and building, environmentalists staked a position that explicitly fought developers for being greedy and ruining the city's sense of place. But their stance also implicitly undermined economic opportunities for blue-collar workers, many of whom were minorities in Austin.

The outcome of these unresolved tensions in environmentalism resulted in an urban planning policy that valued environmental and development outcomes over minority space. When developers, environmentalists, and politicians finally agreed that Austin's environmental amenities were worth protecting, they did not decide to direct growth into the middle- and upper-middle-class neighborhoods that housed many of the city's environmentalists. Rather, they targeted downtown and the city's predominantly minority central eastside for redevelopment. The outcome has been significant displacement of minority residents from neighborhoods that had previously been the only ones available to them, first through institutional policy and then though economic bifurcation. This chapter unpacks these paradoxes and argues that, although mainstream environmentalism in Austin demonstrates an unusual level of citizen agency and grassroots organization, it also reflects the city's troubled racial history and a propensity for environmentalism to exist in tension with racial politics in the contemporary city and suburbs.

THE CITY SUBURBAN

Compared to older cities like Chicago, Philadelphia, or Boston, cities in Texas tend to have much lower population density. The former took shape in eras when people walked or took public transportation and as a result needed to live in close proximity to rail or streetcar lines as well as shops, schools, churches, and sites of recreation. Cities in Texas, and in the South and West more broadly, have grown rapidly in an era when automobiles were common, and the built environment reflects this; because people no longer needed to live close to their daily activities, and because more people could afford single-family homes in the era after World War II, less dense patterns of development proliferated.

Other factors complicate how to think about suburbs, especially in Texas where landscapes tend to be less dense relatively close to the urban core and where cities can take up huge amounts of land (Austin is, for example, about 15 percent larger than Chicago, a city with roughly three times the population). In the most practical sense suburbs are simply distinct municipalities that operate sovereignly but exist in proximity to a large central city. They run themselves through taxation, have independent school systems, governments, and zoning laws, but are considered part of an urban agglomeration anchored by a large city. In another sense we might think of suburbs as a state of mind or set of loosely related social and cultural practices that operate in opposition to "urbanity." People who value peace and quiet, automobile transport, single-family homes, and uniform property values (among many other qualities) might be attracted to certain landscapes, whereas those who prioritize walkability, excitement, and diversity might like others. This is of course a very ideological method of determining what constitutes "suburban." Finally, suburbs can indicate the built aspects of an environment: the character of residences (usually single-family homes with private outdoor space), retail (often malls or strips), zoning laws, and density, regardless of whether a neighborhood exists as part of a large city or as its own separate municipality.

In Austin, defining what counts as suburban is complicated by a unique system of state annexation laws, municipal policies, and demographic shifts that evolved over the second half of the twentieth century. Texas legislators, perpetually interested in stimulating growth, passed the Municipal Annexation Act in 1963, which gave municipalities with more than 5,000 residents the right to annex adjacent smaller territories. Austin itself began annexation policies much earlier; its first annexations came in the 1940s, and in the 1950s the city council voted to enact a generous rebate policy for development on the urban periphery. The postwar era saw robust and consistent growth for Austin; in the 1950s, bolstered by a growing university and state government, as well as nascent technology and leisure sectors, the city's growth remained similar to that of Texas as a whole.[4] By the late 1960s, however, Austin's growth began to outpace most other Texas cities. This was especially true in the home building industry, where Austin's rate of homes built per capita led all large American cities in 1968. The great majority of new homes (as well as shopping malls and some early office parks) were built on the city's periphery, and annexation was key to growth. The city's physical size expanded from 51 to 86 square miles during the 1960s while population density decreased by 17 percent. Accordingly, dependence on the automobile grew quickly as well. Twice as many cars were registered among Austin residents in 1972 as in 1960.[5] So, although 85 percent of Travis County residents lived in Austin in the late 1960s, the city was quite suburban in form with few separate suburbs.

The following decade, however, proved to be pivotal in both how city leaders imagined suburban growth and how environmentally conscious citizens reacted to growth. New ideology regarding urban growth emerged at the state level. In 1975 the Texas legislature created municipal utility districts (MUDs), independent governments that allowed developers to finance infrastructure in new subdivisions by raising bonds. Texas laws also allowed for strong "home rule" powers for cities with more than 5,000 residents, which meant that extraterritorial jurisdiction—the ability of cities to implement zoning, provide services, and broadly govern land use adjacent to their legal boundaries—was practiced widely. MUDs and extraterritorial jurisdiction facilitated peripheral building for developers and annexation for cities by subsidizing infrastructure and coordinating building and zoning laws for new developments. From the city's perspective, annexation was clearly the best choice, given the substantial urban problems facing many older cities whose tax base and middle-class populations were leaving for independent suburbs during the 1970s. Economic decline and stagflation also plagued the United States in the 1970s, and their effects hurt established cities like New York, Pittsburgh, and Detroit acutely. Annexation proved the most effective way to make sure that new subdivisions that would use Austin's resources would also share in residents' burdens by paying taxes and contributing to Austin's economic and social vitality. High-tech firms were an important part of this calculus as well because they generated ample tax revenue for the city. IBM, Motorola, and the city's first publicly traded company, Tracor, all had suburban facilities annexed by the city in the 1970s.[6]

Robust growth in the 1970s gave way to explosive growth in the 1980s as Austin became a national tech hub when Microelectronics and Computer Corporation, a federally sponsored research consortium, picked the city for its location in 1983. In the previous two years, twelve national and global technology and defense firms opened branches in Austin, bringing in roughly 12,000 new jobs, most of which would be filled by knowledge workers who could afford suburban housing. In the early 1980s the city's population growth was among the fastest in the nation. Austin's metropolitan statistical area (MSA) grew from 537,000 to 671,000 in just three years from 1981 to 1985, the largest percentage growth of any U.S. MSA. Demographic growth was again accompanied by spatial expansion, indicating that the same suburban growth patterns were at work. The city grew by over one hundred square miles during the 1980s, an increase of over 80 percent.[7]

So, given the paucity of suburbs in the Austin area from the 1960s to the 1980s, the suburban form of the vast majority of the city, and rapid physical expansion, it makes sense to consider Austin a largely suburban place. Environmentalists who organized in reaction to intensifying growth cited environmental damage from suburban sprawl as the main source of decline, even though the developments they fought, and the natural spaces they sought to protect, were located within the city or adjacent to it.

FROM ENVIRONMENTAL THINKING TO ENVIRONMENTAL ORGANIZING

From its earliest years Austin was in large part defined by its environmental qualities in two senses. One is related to the mix of topography, geology, and weather, the inescapable and often frustrating and dangerous natural circumstances that defined the limits of growth for the city. Water, and often the lack of water, was central. The Colorado River provided the water and picturesque landscape that Mirabeau Lamar used to justify siting the capital there in 1839. In some years, like what Lamar and his cohort described, the Colorado could provide a beautiful and consistent source of surface water.[8] Yet because of its meteorological oddities, Central Texas is prone to some of the worst storms in the continental United States. Storms can come from any direction and, because of the Balcones Uplift directly to the west of Austin, can circle above the city for days. Geologically, the nonporous limestone base that makes up most of the ground in western Austin does not allow water to seep through. U.S. Geological Survey hydrologists consider the area around Austin one of the most flood prone in the United States and have called it "Flash Flood Alley." And though epic floods have devastated the city consistently throughout its history, droughts have been even more commonplace. The city ran out of water dozens of times in the nineteenth century. Thus, Austin's relationship with nature has been strongly adversarial.[9]

The second sense is that Austin's collective vision for its own economic and demographic future, as defined first by business elites and later by diverse groups of people, was also tied to the region's natural environment. Early guidebooks and booster literature listed the city's climate, healthful air, beauty, hilly terrain, and lack of industry as desirable qualities. Boosters consciously differentiated Austin from more industrial cities socially as well; they highlighted the state government and university and downplayed what little industry the city had in an effort to attract higher-class citizens.[10] The Austin Dam, which when completed in 1893 was thought to be the largest in the world, was heralded because it would mitigate the negative effects of flooding, but also because it would attract investment and growth. When that dam was spectacularly destroyed by a flood in 1900, the city sank into a long malaise of low growth. But when the federally funded Highland Lakes dams were completed in the 1930s and 1940s, city leaders immediately began to market Austin as a great place for vacations, knowledge work, and other pursuits that took advantage of the region's natural beauty (and now consistent, safe, and abundant water supply). Nature had been overcome via technology, put to use for humans and then turned into profitable capital in the form of resorts, new suburban-style housing developments, office parks, roads, and strip malls by the 1960s.[11]

Mainstream environmentalism in Austin was born in this context of physical and demographic growth—ironically, growth engendered at least in part by the city's environmental features. Although the movement began slowly, Austinites began to organize against environmental degradation and irresponsible development in the 1960s, and there were instances of citizen agency before that. Many focused on Austin's ample (for Texas) water resources. A spontaneous and successful letter-writing campaign against the privatization of downtown public space along the banks of the Colorado River in 1958 demonstrated a deep interest in access to open space. Writers, many of whom were women, argued that open public space reflected the spirit of the state capital of Texas; commercializing the space, one writer claimed, would "sacrifice this important and long-dedicated open space to the swollen desires of their fellow business men for a fancy meeting place."[12] The Travis County Audubon Society agreed. Its 140 members all signed a letter objecting to a privatized waterfront on the grounds that such a move would undermine the inherently public nature of the state capital.[13]

In the 1960s, Roberta Crenshaw, a west Austin resident, almost single-handedly initiated a sustained campaign against the chamber of commerce, which she saw as a dangerous organization whose only goal was profit via development and privatization. Crenshaw gained control of the city's parks and recreation board (a citizen group, not a city department) in the late 1950s and used it as a platform to organize other environment-minded Austinites. The board fought the city council and chamber of commerce over several issues during the decade, mostly focusing on the city's downtown waterscape, now called Town Lake after the completion of the Longhorn Dam in 1960. Crenshaw began modestly. In 1962 she sought an ordinance to keep motorized craft off Town Lake and to guarantee that public shoreline would stay public. She fought for city funds to enhance vegetation, create walking paths, and enhance undeveloped areas near the water. In 1965, Crenshaw was invited to the White House Conference on Natural Beauty by first lady Ladybird Johnson, who was also beginning efforts to improve Austin's outdoor spaces. Crenshaw used what she learned at the conference to win a federal grant to improve open space around Town Lake in 1968. She was relentless in her efforts to forestall development in public areas that she deemed significant or beautiful; she fought development at Mount Bonnell, a planned expressway on the northern bank of Town Lake, and a proposed amusement park on the south bank.[14]

The movement congealed, however, in response to a threat not from developers or the chamber of commerce but from a University of Texas administrator in October 1969. Under spatial pressure from increased enrollment, administrators sent bulldozers to remove a grove of trees near Waller Creek in the eastern portion of the campus. A group of students met the bulldozers there and forestalled the destruction by occupying the trees and marching in front of bulldozers. Later that week UT regent Frank Erwin, determined to remove the trees, had twenty-seven protesters arrested and instructed workers to remove the trees. His act of aggression, against both nature and peaceful protesters, sparked widespread outrage. Over the following year students organized a massive campaign to get Erwin removed and famously littered the main tower with branches in an action the university's Young Republicans and Young Democrats jointly sponsored to protest the regents' and UT's development policies.[15]

As the city quickly and haphazardly grew in the 1970s, so too did environmentalism, largely in the form of neighborhood organizations. Owing to a pro-annexation policy and liberal annexation laws, the city absorbed an additional forty square miles of land during the decade, about half of its total in 1970.[16] Much of the land was either in the process of being developed or scheduled for development. As a response to citizen concern over the pace of growth and the lack of restrictions on it, in 1973 the city's department of planning created a public planning initiative called Austin Tomorrow, one of the first in the United States that attempted to base a city plan primarily on citizen input. Austin Tomorrow was structured from the ground up, and citizens who participated met with people in their neighborhoods to discuss all types of urban issues. The plan that was adopted emphasized environmental as well as social concerns, but it had no legal backing and was never implemented. Yet, although the document had less relevance than proponents had hoped, the process encouraged grassroots local organizations, which connected and energized many citizens around environmental issues. The number of neighborhood-based groups more than doubled between 1973 and 1980, from twenty-nine to sixty-six; by 2000 nearly two hundred neighborhood environmental groups covered most of Austin and many suburbs.[17]

In response to even more intense growth, in the 1980s Austin's disparate neighborhood groups increasingly saw commonality in their issues and interests, and they often expanded out of their neighborhoods to form larger coalitions with broadly conceived goals. The nature of Austin's physical environment and the growing cadre of national developers building in and around the city provided threats that linked numerous Austinites and suburbanites together. The creation and extension of MOPAC highway, which cut a 300-foot-wide swath through west Austin, was one of the first issues to bring environmentalists together from myriad neighborhoods. Roads, and specifically large highways, provided a unique threat to environmentalists. From a scientific standpoint roads created impervious cover, pavement that repelled water after mixing it with oil and other chemicals left on the roads. Such water has fewer opportunities to reach the rivers, creeks, and aquifers in a pure form. From a social perspective, highways brought ugly billboards and carved up hundreds of acres of undeveloped land, eviscerating existing neighborhoods.[18]

But perhaps more than anything, highways allowed the city and suburbs to expand. Highways, and MOPAC in particular, with its reach into the sensitive areas above the Edwards Aquifer in southwest Austin, directed growth. Developments became more profitable and more common in areas serviced by highways. Motorola, for example, opened a large office over the Edwards Aquifer after MOPAC was finished. MOPAC also had a binding effect for geographically distant environmentalists because it connected otherwise separate places. Most of the residential and retail developments that engendered citywide concern were serviced by new highways. The Barton Creek Square Mall and the Circle C subdivision, both of which brought diverse environmentalists together, were made possible by the MOPAC extension. Finally, the highway pitted residents of older, more central parts of the city against new suburban developments and their residents. Why should we have to subsidize growth, residents argued, in areas that we do not want to see developed?[19]

Although MOPAC linked distant areas and environmentalists from across the area, Austin's waterways, as they had in the past, became the central symbol of the threat posed by irresponsible development. Like highways, creeks and rivers also link distant places together; but threats to the city's water supply and recreational areas had even more existential meaning. Those threats, most often, were directly related to development. Early in the 1980s the Zilker Park Posse, one of Austin's first environmental interest groups, was able to link construction of Barton Creek Square Mall with the declining condition of Barton Springs Pool, the most famous natural area in the city. The pool was shuttered thirty-two times in 1981, leading the Posse to advocate for hydrological studies of the impact of construction and eventually to run television commercials linking construction and Barton Springs.[20]

Throughout the 1980s, Barton Springs, and the increasing threats posed to it, became the central symbol of Austin's environmental movement. In the early 1980s, the Posse and other neighborhood groups helped to write many low-density ordinances for areas around Barton Creek and other urban creeks, including the Barton Creek Ordinance, one of the nation's strongest water quality ordinances at the time. In 1985 environmentalists helped to pass a comprehensive watershed ordinance intended to regulate the amount of runoff that developers could allow into Austin's creeks across most of the city. In 1985 environmentalists also won an important political contest when Frank Cooksey beat development-friendly incumbent Ron Mullen in the city's mayoral race. Yet these victories proved empty in practical terms. Developers were able to file for zoning variances to sidestep water quality ordinances; 80 percent of new developments were granted variances in the early 1980s.[21] Austin saw record population growth and physical expansion. The city still had no legal jurisdiction over MUDs, which remained the development style of choice and were easily annexed into the city. By 1987, when the Texas legislature passed a bill forbidding cities from changing water quality laws after a development was begun, environmentalism in Austin appeared vulnerable.

Yet, perhaps fittingly, the 1987 economic downturn and real estate bust provided new momentum for environmentalists. During that year, 4,400 Travis County homes were foreclosed upon, and roughly the same number of commercial properties were taken over by banks and lending institutions. Two of the largest local developers, including Gary Bradley of Circle C Ranch, filed for bankruptcy. Office space for rent around the city dropped in value by half from 1986 to 1987; by December, Austin had the highest vacancy rates for office space in the entire country. The robust building spree and optimistic economic outlook of the early 1980s ground to an immediate halt.[22] Global conglomerate Freeport McMoRan entered the fray in 1988. The corporation immediately purchased 3,000 acres of land above the Edwards Aquifer, loaned Gary Bradley the money to finish Circle C, and sold the Barton Creek Country Club to a national country club management corporation. President Jim Bob Moffett then announced plans to build the largest ever development in Austin, the Barton Creek Planned Unit Development, a 4,400-acre multiuse community with 2,500 homes, 1,900 apartment units, 3.3 million square feet of office and retail space, and four golf courses, all overlooking Barton Creek.[23]

The environmental threat posed by Freeport McMoRan and Moffett was obvious: high density, asphalt, construction, wastewater and refuse, and traffic over Edwards Aquifer, not to mention the loss of open space. But for many the threat was existential and struck at the heart of Austin's authentic identity. Globally, Freeport was a multinational corporation known for rampant environmental and human rights abuses, particularly with mining. To environmentalists, Freeport did not care what happened to Austin's sense of place. They were outsiders who stood to profit in the short term without giving thought to the community's long-term health and happiness; because of their complicity with the city council, many of Austin's politicians were considered so as well. Signs of environmental decline were already ubiquitous. Barton Springs pool continue to close under mysterious circumstances. The size and population of the city continued to grow. And by 1990 two species unique to the Barton Creek watershed were labeled as endangered.[24]

Austin's environmental movement, and the sacred place of Barton Springs in the city's environmental mythology, coalesced in the summer of 1990 as the city council prepared to determine the fate of the Barton Creek Planned Unit Development. Hundreds of people, many of whom had little or no previous contact with environmentalism in Austin, gathered at Barton Springs the night before the hearing for an impromptu vigil. Hundreds more gathered outside the council chamber the day of the hearing, waving signs and encouraging passing cars to "take back our city." More than seven hundred citizens signed up to speak at the hearing, which quickly turned into a sort of variety show, with poetry readings, home videos, and golf balls pulled from the creek dumped on the chamber floor. Jim Bob Moffett was heckled mercilessly. Barton Springs defenders spoke until 6 A.M., when a weary council voted unanimously to disallow the development.[25]

The Barton Creek development hearing became the defining moment for environmentalists in Austin because it demonstrated that grassroots organization, and residents' passion for both Barton Springs and the greater idea that Austin's sense of place was largely defined by its natural environment, could defeat entrenched developers and politicians. They used the hearing as a point from which to create a systematic defense of Barton Springs. In the next few years, activists pressured the Texas Water Commission to protect the springs from pollution. Their new group, the Save Our Springs Alliance, wrote a new, more powerful ordinance and collected 30,000 signatures supporting it. The ordinance outlawed zoning variances in protected areas, limited impervious cover, mandated water quality testing, and limited density throughout the watershed.[26]

THE MANY USES OF DENSITY

During the 1970s and 1980s the demographic and physical composition of peripheral neighborhoods was imperative. For many neighborhood groups, keeping apartment complexes and retail stores out of their single-family neighborhoods was foremost among their concerns. In the 1960s and 1970s, three trends conspired to increase the number of renters in Austin. First, as a result of increasing population and growing funding for universities, University of Texas enrollment increased dramatically. The university, which found expansion difficult because of its central location, made it easier for students to live off campus and radically reorganized the city's bus system to transport students from around the city. Second, when the university was able to expand, it employed eminent domain under federal urban renewal law to do so. The City of Austin participated in the program as well. Overall, more than one thousand homes were destroyed, and evicted residents, many of whom were poor, were forced to enter the rental market.[27] Finally, rapid population increase meant that many new-comers had to rent upon moving to the city. Taken together, the number of renters in the city soared in the 1970s, and the trend continued into the 1980s as well.

Older, more environmentally focused groups tended to see higher den-sity as a primarily environmental concern and attacked new developments on environmental grounds. This perspective emphasized impervious cover and the polluting aspects of construction as well as potential damage caused by building in sensitive areas near creeks. Some groups pointed to a 1975 study of the Barton Creek watershed that found that increasing imper-vious cover had deleterious effects on the creek and adjacent land.[28] In the late 1970s, for example, the Zilker Park Posse pointed to higher-density developments in the northwest as the main cause of Lake Austin pollution via Bull Creek, one of its largest tributaries.[29] Environmentalist and lawyer Bill Bunch and others argued that the city should enforce strict building guidelines over the Edwards Aquifer Recharge Zone because impervious cover there would destroy Barton Springs; low-density developments were the only way to mitigate environmental disaster for the Springs.[30] For others, the increased traffic would create more pollution that would even-tually make its way into the aquifer.[31] Like apartment complexes, shopping centers created large amounts of impervious cover because they catered so completely to automobiles.

Though many groups were more environmentally focused, they folded environmental issues into broader, localized concerns about quality of life. For some, environment and quality of life became difficult to dis-tinguish. Density was often at the heart of their arguments, and almost always zoning was key as well. Strip malls and stores were disparaged. Allowing for retail zoning not only would spoil the view of the cliffs along Barton Creek's banks, it would also expose loading docks, garbage dump-sters, and parking lots to hikers and residents. The Posse worried about "commercial degradation of the scenic beauty surrounding Zilker Park." In Rollingwood, a highly suburban neighborhood west of Zilker Park, residents complained about the potential "visual pollution" and "strip development" and garbage dumpsters behind stores that new retail zoning would force onto the community.[32] In a likely attempt to reach developers, some citizens attempted to cast Austin's pleasing environment as an economic advantage for the city. "Quality supplies of water and recreational opportunities in close proximity to Austin" were essential for attracting growth because they made the city desirable in the first place.[33]

For some vociferous citizens and neighborhood groups, less-restrictive zoning and higher-density development caused social consternation. Neighborhood group United South Austin feared "overzoning" in their area and linked it with apartment complexes and shopping centers that caused flooding and would necessitate a new, and undesirable, wastewater treatment facility.[34] The Travis Audubon Society wrote to the planning commission in 1978 to argue against zoning changes that it claimed would destroy south Austin's neighborhood fabric. For them the main threat was Austin becoming another "apartment city," and they pointed to East Riverside, a predominantly Latino area, as proof of the negative effects of less-restrictive zoning laws.[35] In 1979 the Zilker Park Posse and the Sierra Club requested an apartment moratorium for areas around Barton Springs. Apartment buildings, though bad for the environment, also detracted from Austin's unique sense of place. "Austin is a unique and wonderful city," said state representative Mary Jean Bode, and the relationship between the community and the natural environment "is one of the reasons it's so unique."[36] In Great Hills Trail, a new subdivision on Austin's northwest periphery, hundreds of residents wrote to voice their displeasure for a new office building and the zoning that allowed for it.[37] To them and many others the word "residential" became a rhetorical device indicating distaste for any addition to the physical landscape other than single-family homes, parks, and civic buildings. To many environmentalists and other concerned citizens, apartment complexes did not reflect Austin's sense of place or their own aspirations for their city.

Perhaps the most vociferous citizens saw apartments as blights on their neighborhoods' social world and as economic penalties for homeowners. Residents from northwest Austin especially saw apartments in this way, and their letters indicated that they were set to battle them. Writers from the Walnut Creek area believed that apartments increased crime and congestion as well as undermining the area's environmental amenities and sense of place. "Please do not give us another Houston!" remained a common sentiment among Austinites fighting specific density increases as well as general growth.[38] Some flatly indicated that growth should take place elsewhere, although there was no indication where that place might be.[39] The most forceful anti-apartment comments linked apartment dwellers with a lack of morality and characterized the owners of single-family homes as virtuous citizens being put upon by "transient occupancy." Apartments forced homeowners to engage in constant "neighborhood upkeep" and destroyed "neighborhood atmosphere." Economically, homeowners were "penalized" via stagnant or decreasing property values ushered in by apartments and the "trash and vermin" who lived there. One resident summed up his feelings quite plainly, writing, "No homeowner raising a family wants to bring up children across the street from 20 acres of medium density apartment complexes." Another pointed to the civic responsibility that renters lacked, writing, "We *care for* and take pride and interest in our homes because we *own*, not rent." Here, density is conflated with multiple social and economic maladies that promised neighborhoods would "sink under the weight of apartment complexes."[40]

Evidence also exists that neighborhoods that were less wealthy, less white, and not on the city's westside were more often cited as appropriate places to increase density. When the city approved federally sponsored housing units in the late 1970s, it targeted areas that were already denser and less white than average in the city. Members of South Austin Neighborhood East (SANE) fought the units aggressively, arguing that theirs was an already integrated neighborhood and that the subsidized apartments should be located in areas with a larger percentage of white residents. West Austin, with numerous neighborhoods housing well over 90 percent white residents, was an obvious target of SANE's argument.[41]

The residents who voiced their opinions about the negative effects of higher density came from many neighborhoods and provided many different perspectives. For some, environmental issues were primary. Others saw the potential social and economic deterioration as the greatest threat. Still others worried about their neighborhood's quality of life, aesthetic attributes, and traffic. Yet what bound them all together was an aversion to increases in density in the neighborhoods they lived in and near the areas they wanted to protect. In part distaste for increased density was the outcome of Austin Tomorrow, which found that more intense uses of older central areas led to less desirable outcomes and the flight of residents. Neighborhood groups that emerged from Austin Tomorrow already looked at higher density as a problem to be avoided.[42] Yet for some residents density became an existential problem that linked apartments with myriad forms of social, environmental, and economic decline. In many more sub-urban westside neighborhoods the aversion to higher density went back decades, codified in restrictive covenants beginning early in the twentieth century.[43] By the late 1990s, as the city embarked on a radically new urban planning philosophy generated by the success of mainstream environmen-talism, the geography of development shifted to the central city.

DENSITY, DISPLACEMENT, AND THE CITY

In the late 1990s, Austin voters elected the "Green Council"; all seven city council representatives had ties to the environmental movement. Mayor Kirk Watson quickly created Austin's EPA-endorsed Smart Growth Ini-tiative, an environmentally inspired urban planning philosophy that drew on the tenets of "new urbanism": higher density, walkability, aggregate housing, land development boundaries, public transportation, all facili-tated by multiple-use zoning. Importantly, Watson and the Green Council also recast the role of environmentally responsible development in urban growth. For decades environmentalism and growth were seen as polar opposites. With Smart Growth, Watson argued, protecting the environ-ment could make Austin more attractive for investment and development. His idea proved almost universally admired. The Austin Real Estate Coun-cil, the chamber of commerce, and the leading environmental group, Save Our Springs Alliance, all supported the plan, which essentially sought to incentivize development in central areas and set aside undeveloped land along the city's environmentally sensitive western perimeter. Smart Growth in Austin was thus seen as a way to preserve the city's beauty, its recreational sites, and, for some, its sense of place while still facilitating demographic growth and the profits that came with it.[44]

Implementation of the Smart Growth Initiative was swift and effec-tive. In 1998 the council created two bond packages worth close to $1 billion dollars to save undeveloped land on the urban fringes (mostly in west Austin) and promote new development in the existing urban core. Voters passed them. At the same time, the city approved widespread zoning changes in urban areas marked for investment that would increase density standards and pave the way for new urban–style developments as well as centrally located high-tech firms and other businesses. A frenzy of construction occurred in the central business district over the next two years as tech businesses and apartment complexes signed up to relocate to the urban core. Most residential neighborhoods, especially those on the westside, remained similar in demographic composition and density. Growth continued, but the geography of growth shifted radically—both to the central core and to fast-growing, less affluent suburbs predominantly north of the city.[45]

The people living in the central core, and particularly the minority communities who had been forced to live in central east Austin for most of the twentieth century, were left out of the equation. As money and investment poured into their neighborhoods, places that discriminatory zoning and real estate policy had kept from prospering for most of the twentieth century, prices and taxes increased drastically. New urbanist–inspired residential-retail developments began to dot the landscape as zoning changes encouraged mixed-use development. Gentrification followed. Eastside environmental groups, who fought for cleaner, healthier, and more equitable minority space in the 1990s, attacked gentrification to little avail.[46] In 2002 the Travis County appraiser announced that prices in east Austin were increasing faster than anywhere else in the nation. Land values increased by 400 percent in East Cesar Chavez, a predominantly Latino neighborhood adjacent to downtown, from 1998 to 2004. That year the Austin Human Rights Commission found the situation so troubling that it recommended a ninety-day construction moratorium in East Cesar Chavez.[47] North of there, in the predominantly African American neighborhoods along Eleventh and Twelfth streets, development came slower. Yet from 2000 to 2010 all four census tracts between Seventh Street, Manor Road, and Interstate 35 lost 15.6 percent of their African American population; those neighborhoods also saw an increase of roughly 20 percent in their white population. Ironically, environmentally sustainable development has often proven decidedly unsustainable for many of Austin's minority residents. Though Austin has lost African American population share every decade since 1920, it is currently the only large fast-growing city in the United States that lost aggregate African American population between 2000 and 2010, even as its overall population increased by almost 20 percent.[48]

The displacement of minorities from east Austin has also led to interesting trends for Austin's fast-growing independent suburbs, which have exploded since the 1990s. Taken together, Round Rock, Pflugerville, Leander, Cedar Park, Hutto, and Georgetown, six large northern suburbs, grew from approximately 60,000 residents in 1990 to 140,000 in 2000 and to 375,000 in 2015. Many of those residents are minorities. Austin's metro region is now one of the few in the country that has a higher percentage of African Americans living in the suburbs than in the city.[49]

As Austin's urban planning policies begin to focus on the central city, the demographics and meaning of Austin's suburbs change. In the 1970s and 1980s the city was eager to annex new subdivisions, extend the municipal boundaries of the city, and capture the tax revenues generated by upper-middle-class homes, retail shopping, and technologically oriented businesses. Today the geography has changed. The city appears less interested in physical expansion and more interested in promoting itself as an environmentally responsible, sustainable city. The sharp increase in land values, and property taxes, has proven this ideology profitable for the city. But sustainable policy has perhaps been less sustainable for more disadvantaged Austinites and for the suburbs to which they increasingly migrate. Whereas the city was once a more diverse place socially and economically, today the diversity of the suburbs is increasing. At the same time, poverty is rapidly increasing in Austin's suburbs, up 143 percent from 2000 to 2011; this increase is one of the sharpest among all U.S. metro areas.[50]

NOTES

1. *JW Properties* lists many of Austin's environmental awards and green initiatives at http://jwproperties.net/default.asp.pg-AustinaGreenCity (accessed September 15, 2015). For trails, see "The Best Fitness Walking Cities," *Prevention* 59, no. 4 (April 2007): 118, http://content.time.com/time/magazine/article/0,9171,2103780,00.html (accessed July 27, 2015); and Steven A. Moore, *Alternative Routes to the Sustainable City: Austin, Curitiba, and Frankfort* (Lanham, MD: Lexington Books, 2007). For a fairly comprehensive list of what the city has done and is planning, see Howard Witt, "Austin's Green Ambitions," *Seattle Times*, October 5, 2007.

2. Paul Burka, "The Battle for Barton Springs," *Texas Monthly* (August 1990), 74.

3. William Scott Swearingen, *Environmental City: People, Place, Politics, and the Meaning of Modern Austin* (Austin: University of Texas Press, 2010); Richard White, "'Are You an Environmentalist or Do You Work for a Living?': Work and Nature," in William Cronon, ed., *Uncommon Ground: Rethinking the Human Place in Nature* (New York: W. W. Norton, 1996), 171–85; Joseph E. Taylor III and Matthew Klingle, "Environmentalism's Elite Tinge Has Roots in Movement's History," *Grist*, March 9, 2006, http://grist.org/article/klingle (accessed May 25, 2016).

4. See, for example, Vic Mathias, "Should Our Growth Continue?," *Austin in Action* 2, no. 8 (January 1961): 14–17; Austin Department of Planning, "Basic Data about Austin and Travis County" (Report, 1955), Vertical File "Austin, Texas—City Planning (I)," Briscoe Center for American History, University of Texas at Austin (hereafter BCAH); see also Andrew M. Busch, *City in a Garden: Environmental Transformations and Racial Justice in Twentieth-Century Austin, Texas* (Chapel Hill: University of North Carolina Press, 2017), chaps. 4 and 5.

5. Larry BeSaw, "Building Permits Breaking Records," *Austin American*, September 6, 1967, Folder "General Texas Austin 1968," Box 95–112–203, Papers of J.J. Pickle, BCAH; "Austin Building Sixteenth in Nation," *Austin Home Builder*, March 7, 1968, Folder "General Texas Austin 1968," Box 95–112–203, Papers of J. J. Pickle, BCAH; Larry BeSaw, "Apartments, Autos Expand Role in Austin's Lifestyle," *Austin American*, September 6, 1973.

6. Swearingen, *Environmental City*, 92–94; League of Women Voters, "The Economics of Annexation," unpublished report, April, 1984, Folder "3–16," Box 3, Mather (Jean) Papers, Austin History Center, Austin, Texas (hereafter AHC).

7. *Imagine Austin Comprehensive Plan* (Austin: City of Austin, 2012), 32.

8. Jeffrey Stuart Kerr, *Seat of Empire: The Embattled Birth of Austin, Texas* (Lubbock: Texas Tech University Press, 2013), 8–10; Eugene C. Barker, "Description of Texas by Stephen F. Austin," *Southwestern Historical Quarterly* 29 (October 1924): 98–121.

9. Terry Jordan, *Texas: A Geography* (Boulder, CO: Westview Press, 1984); E. T. Dumble, *Second Annual Report of the Geological Survey of Texas* (Austin: State Printing Office, 1890); Student Geology Society, "Guidebook to the Geology of Travis County," report, University of Texas at Austin, 1977; Jonathan Burnett, *Flash Floods in Texas* (College Station: Texas A&M University Press, 2008). The city has run out of water numerous times, as late as 1885. "'Austin Dam' Report—Notes on Water and Light History Minute Book Entries 1884–1922, AHC; "Brief History of the Austin Dam," Folder "Austin Dam Correspondence 1938," Box 5, Office of the City Clerk Records, AHC.

10. "Austin and Travis County, Texas: Charms of the Capital City and Its Environs: Attractiveness and Resources of Texas and Peculiar Organic Code of the Great Commonwealth" (Austin: Democratic Statesman and Book and Job Stream Print, 1877); and "Austin, the Capitol of Texas, and Travis Country" (Austin: Dupre and Peacock, 1876), both in BCAH; American National Bank, "A Citizen of No Mean City" (pamphlet, 1922); Austin Commercial Club, *Austin, Texas, The Future Great Manufacturing Center of the South. The Healthiest City in the South. Facts for Consideration of Tourists, Home-Seekers, Investors, Manufacturers, and Merchants* (Austin: Ben C. Jones, 1891).

11. Busch, *City in a Garden*, 86–107.

12. Mrs. R. Q. Underwood to Austin City Council, November 18, 1957, Box "Robert Thomas Miller, April 1956–December 1957," Tom Miller Papers, AHC.

13. Richard G. Underwood to Hon. Charles F. Herring, November 16, 1957, Box "Robert Thomas Miller, April 1956–December 1957," Tom Miller Papers, AHC.

14. These documents, from Box 1, Crenshaw (Roberta) Papers, AHC, illustrate support for anti-development measures: Roberta P. Dickson to Mr. Walter Wendlandt, March 1, 1968, Folder "Town Lake Policy Zoning"; (Mrs.) Roberta P. Dickson to Mr. Stuart King and Mr. Alan Y. Taneguchi," December 7, 1962, Folder "Park Board—Town Lake"; "Triple Award Goes to Mrs. Dickson," *Austin American*, January 21, 1972, loose in box; "Speights" 7-7-68, Folder "Town Lake Concept PARD"; and Roberta P. Dickson to Mayor Lester Palmer, June 25, 1965, Folder "PARD Board." Citizens other than Crenshaw supporting anti-development measures: Irving R. Ravel to Mrs. Fagan Dickson, November 30, 1964, Folder "PARD Board"; Henrietta Jacobsen to the Honorable Lester Palmer, October 23, 1966, Folder "PARD Board."

15. Jenny Rice, *Distant Publics: Development Rhetoric and the Subject of Crisis* (Pittsburgh, PA: University of Pittsburgh Press, 2012), 1–3, describes the incident; Ed Smith, "'Ax Erwin' Rally Attended by 300," November 5, 1969, Folder "Waller Creek," Box 1, Crenshaw (Roberta) Papers, AHC.

16. Austin City Council, *Imagine Austin Comprehensive Plan* (Austin: City of Austin, 2012), 32.

17. Swearingen, *Environmental City*, 104.

18. "Maureen McReynolds to Dick Lillie," June 22, 1978, Folder "105 Zoning Changes in Barton Creek Area 1978–1983," Box 1, Zilker Park Posse Records, AHC.

19. Swearingen, *Environmental City*, 116, 121–25, 142–47.

20. Burka, "Battle for Barton Springs," 74.

21. Swearingen, *Environmental City*, 143–48.

22. Kim Tyson, "Real Estate Take a Fall for the Year," *Austin American Statesman*, December 27, 1987, Vertical File "Austin, Texas—Housing and Real Estate (Travis County)," BCAH.

23. Burka, "Battle for Barton Springs," 72; Swearingen, *Environmental City*, 147–48.

24. Burka, "Battle for Barton Springs."

25. Swearingen, *Environmental City*, 148–51.

26. Andrew Karvonen, *Politics of Urban Runoff: Nature, Technology, and the Sustainable City* (Cambridge, MA: MIT Press, 2011), 53–57.

27. Eliot Tretter, *Shadows of a Sunbelt City: The Environment, Racism, and the Knowledge Economy in Austin* (Athens: University of Georgia Press, 2016), 33–56;

Andrew M. Busch, "Building 'A City of Upper-Middle Class Citizens,'" *Journal of Urban History* 39, no. 5 (September 2013): 975–96.

28. Swearingen, *Environmental City*, 121–23.

29. "Reelect Ron Mullen, "Folder "1/4," Box 1, Zilker Park Posse Records, AHC.

30. Bill Bunch, "Securing a Safe Future for Barton Springs: A Position Paper" (unpublished manuscript), Folder 1, Box 1, We Care Austin Records, AHC; "Maureen McReynolds to Dick Lillie," June 22, 1978, Folder "105 Zoning Changes in Barton Creek Area 1978–1983," Box 1, Zilker Park Posse Records, AHC.

31. Petition, Folder "Zoning Changes in Barton Creek Area 1978–1983," Folder 1/1, Box 1, Zilker Park Posse Records, AHC; Phillip S. Blackerby and Alexandra K. Blackerby, testimony, Folder "105 Zoning Changes in Barton Creek Area 1978–1983," Box 1, Zilker Park Posse Records, AHC.

32. Phillip S. Blackerby and Alexandra K. Blackerby, testimony, Folder "105 Zoning Changes in Barton Creek Area 1978–1983," Box 1, Zilker Park Posse Records, AHC.

33. Ordinance No. 80 4/1/80, Folder 1/2, Box 1, Zilker Park Posse Records, AHC.

34. United South Austin, "Meeting Notice and Agenda," November 30, 1984, Folder 1/7, Box 1, Arnold (Mary) Papers, AHC.

35. "Mike Thomasson to Members of the Planning Commission," August 29, 1978, Folder "Zoning Change in Barton Creek Area 1978–1983," Box 1, Zilker Park Posse Records, AHC.

36. Bill Collier, "Zilker Posse to Demand Apartment Moratorium," *Austin American Statesman*, June 19, 1979; Terry Peters, "Posse Urges Barton Moratorium," *Austin Citizen*, June 12, 1979.

37. See the dozens of letters in Folder 1/16, Box 1, Arnold (Mary) Papers, AHC.

38. Brenda M. Berger to Ms. Mary Arnold, February 27, 1985, Folder 3/3, Box 3, Arnold (Mary) Papers, AHC; "Georgia F. Horn to Planning Commission Members," February 20, 1985, Folder 3/3, Box 3, Arnold (Mary) Papers, AHC.

39. "Ginger Simmons to Planning Commission Members" (n.d.), Folder 3/3, Box 3, Arnold (Mary) Papers, AHC.

40. Most quotes here are taken anonymously from a document titled "Comments," although one person did give his name. See "Comments," Folder 2/11, Box 2, Arnold (Mary) Papers, AHC; see also "Marc and Deanna Warner to Planning Commission Members," February 20, 1985, Folder 3/3, Box 3, Arnold (Mary) Papers, AHC.

41. Untitled document dated February 26, 1977, in folder "Save Austin Neighborhood East, 1977," Folder 3/7, Box 3, Mather (Jean) Papers, AHC; "Project's Okay Ires Residents," *Austin Citizen*, February 25, 1977.

42. See, for example, "1983—W. Austin Neighborhood Association," Folder 3/11, Box 3, Mather (Jean) Papers, AHC.

43. Eliot Tretter and M. Anwar Sounny-Slitine, *Austin Restricted: Progressivism, Zoning, Private Racial Covenants, and the Making of a Segregated City* (Austin: Institute for Urban Policy Research and Analysis, 2012).

44. For smart growth and new urbanism broadly, see Andres Duany, Jeff Speck, with Mike Lydon, *The Smart Growth Manual* (New York: McGraw-Hill, 2010); and Aaron Passell, *Building the New Urbanism: Places, Professions, and Profits in the American Metropolitan Landscape* (New York: Routledge, 2013). For Austin's smart growth plan and other urban planning ideology, see Greater Austin Chamber of Commerce, *Next Century Economy: Sustaining the Austin Region's Economic Advantage in*

the 21st Century (Austin: Greater Austin Chamber of Commerce, 1997); Joel Warren Barna, "The Rise and Fall of Smart Growth in Austin," *Cite* 53 (Spring 2002): 22–25; and Mike Clark-Madison, "A City with Smarts: Austin Wising Up to Growth Plans," *Austin Chronicle*, April 17, 1998, www.austinchronicle.com/news/1998–04–17/523318 (accessed July 20, 2011).

45. Carl Grodach, "Before and after the Creative City: The Politics of Urban Cultural Policy in Austin, Texas," *Journal of Urban Affairs* 34, no. 1 (2012): 81–97; Barna, "Rise and Fall." Austin's north suburbs have grown vigorously since 1990.

46. Tretter, *Shadows of a Sunbelt City*, 96–113; Busch, *City in a Garden*, 226–51.

47. PODER, "SMART Growth, Historic Zoning and Gentrification of East Austin: Continued Relocation of Native People from their Homeland," Folder "Gentrification Report 2003," Box 33, PODER Records, AHC; "East Austin Housing Forum," Folder "PODER Housing Forum 2006," Box 32, PODER Records, AHC. The University of Texas School of Architecture and Community and Regional Planning did the study of land values and property taxes; Welles Dunbar, "How Not to Gentrify: HRC Asks for Eastside Moratorium," *Austin Chronicle*, November 4, 2005.

48. City of Austin, "Tract Level Change, 2000 to 2010, Total Population, Race and Ethnicity," spreadsheet, *Austintexas.gov*, www.austintexas.gov/page/demographic -data (accessed March 18, 2015); Eric Tang and Chunhui Ren, *Outlier: The Case of Austin's Declining African American Population* (Austin: Institute for Urban Policy and Research Analysis, 2014).

49. Elizabeth Kneebone and Alan Berube, *Confronting Suburban Poverty in America* (Washington DC: Brookings Institution, 2014).

50. Kneebone and Berube, *Confronting Suburban Poverty.*

THE FORGING OF AN AFRICAN AMERICAN COMMUNITY ON THE OUTSKIRTS OF THE ALAMO CITY, 1980–2010

HERBERT G. RUFFIN II

To date few scholars have explored African Americans in San Antonio, let alone black suburbanization. Authors who have published on black San Antonio include Alwyn Barr, Larry P. Knight, Robert C. Fink, Kenneth Mason, Jerelyne Castleberry Williams, Judith Kaaz Doyle, and Robert A. Goldberg. The topics these authors have researched range from race relations during Reconstruction and the Jim Crow era to black urban churches in South Texas, semiprofessional African American baseball, education and empowerment, and civil rights activism.[1] Informed by literature on black migration and community formation in twentieth-century urban and suburban America, especially the work of Albert S. Broussard, Sheryl Cashin, Lawrence De Graaf, Karyn Lacy, Thomas Sugrue, Quintard Taylor, Joe William Trotter, and Andrew Weise, this chapter extends the chronology and scope of black San Antonio scholarship by examining African American empowerment in north San Antonio from 1980 to 2010.[2] In particular, it addresses tactics that black suburban San Antonians have implemented to empower themselves—namely, through migration, suburbanization, and community formation—and ends with the question of whether or not class division prevents black San Antonio suburbanites from rebuilding a sense of community with African Americans in the city's eastside, San Antonio's predominantly black working-class neighborhood. I begin by setting the black suburbanization foundation with discussions of San Antonio's twentieth-century urban development story, Jim Crow's impact on the city's African Americans, and how civil rights legislation has sparked black migration away from the central city.

THE ALAMO CITY AS A TRIFURCATED SPACE

Since 1980, San Antonio has developed into a trifurcated city, which means that it has been socioeconomically and geographically divided into three distinct urban communities.[3] This divide has occurred along the lines of big-city rapid growth in the northern suburbs, stagnated growth in west, south, and east San Antonio (which are populated by African Americans and Mexican Americans), and gentrification of the Alamo-Riverwalk section downtown.

Key to understanding San Antonio's urban divide is its freeway system. Since 1961, San Antonio's urban growth has been developing around its beltway, Interstate 410. After 1977 this development became rapid suburban growth with the development of a ninety-mile beltway system called Loop 1604 that wraps around San Antonio and the I-410.[4] Connecting these two beltways is a series of freeways that run through the city—U.S. Highway 281 North and Interstates 10 and 35 in north San Antonio. Similar to other postwar southwestern cities, such as San Jose and San Diego, most of San Antonio's suburbanization exists within cities and communities annexed within its city limits. Only 20 percent of San Antonio's suburbs exist beyond its city limits.[5]

To date, the most developed suburban area is the "Golden Triangle," encompassed within the I-410, Loop 1604, I-10, and US 281. This area has been the home to some of the most white and affluent communities in San Antonio and in incorporated cities surrounded by the Alamo City. More important, it is this area that has rapidly grown into San Antonio's first modern suburb, resulting in it being the city's model for contemporary growth. Prior to 2000, these suburbs stereotypically included the affluent white enclaves of Castle Hills, Churchill Estates, Hill Country, Hollywood Park, Shavano Park, and Vance Jackson. Today they mostly exist as gated communities beyond I-410 and have been expanding north of Loop 1604 to Medina, Bandera, Kendall, Comal, and Guadalupe counties.[6]

The second distinct group of communities exist within the I-410, which is considered by many locals to be the city center.[7] Prior to the development of I-410 in 1961, incorporated cities above Hildebrand Avenue, such as Alamo Heights, Olmos Park, and Terrell Hills, were considered suburbs of San Antonio's city center. This center was the Main Plaza surrounded by thirty-six square miles.[8] In the 1920s and 1930s, these towns formed with the advent of the automobile, and with affluent white families "escaping" the perceived economic, social, and political ills of city life while holding on to political economic power closely associated with San Antonio. Initially Alamo Heights, Olmos Park, and Terrell Hills restricted people of color with restrictive deeds. After restrictive covenants were federally outlawed in 1948, through the U.S. Supreme Court verdict *Shelley v. Kramer,* according to historian Frank Char Miller, San Antonio suburban communities became "built to serve as tax havens for the commercial elite, and they wanted to keep them racially and economically segregated." To date, all of these communities still are predominantly white and affluent, which inspired Miller to state that, "No, the 'wall' isn't there, but it tells you how powerful the class rules were when you don't have the covenants any longer but you still have much of the same demographics."[9]

After I-410 was built in 1961, Alamo Heights, Olmos Park, and Terrell Hills became high-property-value incorporated communities surrounded by San Antonio—which has been annexing communities and open space toward the city's Los Angeles Heights and North Loop communities since 1945.[10] Other incorporated spaces that have followed their examples and exist on the northern fringe of the Alamo City include Balcones Heights, Castle Hills, Hill Country Village, Hollywood Park, Kirby, Leon Valley, Shavano Park, Terrell Hills, and Windcrest. Since 1980, northeastern suburbs between the I-410, Loop 1604, US 281, and I-10—which include the incorporated towns of Converse, Kirby, Universal City, and Windcrest—have been areas where black populations ranging from 5 to 30 percent have increasingly made their homes.[11]

Finally, San Antonio's third distinct community is the revitalized Alamo-Riverwalk area. Here San Antonians live in mixed land-use communities that combine residential, industrial, and commercial land uses all within the same acreage, unlike the suburbs to the north. In this sense, it is similar to other U.S. metropolises. Part of this post-1968 HemisFair (San Antonio World's Fair) growth included the mass development of condos, lofts, and revitalized historical homes for urban dwellers and young professionals not ready to settle down with families in suburbia.[12] Since 2000, a small percentage of recent black college graduates who work in the city have also participated in this growth pattern, which extends from the Riverwalk area to southern downtown historical districts near the San Antonio River like King Williams.[13]

SAN ANTONIO'S COLOR LINE

Since the 1940s, San Antonio's city center has been peculiarly divided along the lines of race.[14] In this period, the eastside has been known as the city's black community. Though the westside has always housed about 15 percent of African Americans, this area as well as the southside are predominately Mexican American communities with populations ranging from 75 to 95 percent.[15] During the postwar period (1945–68), black San Antonians were confronted with racial discriminatory barriers in housing, employment, education, and police misconduct.[16] Although this was the case, most African Americans born in the 1930s to 1950s saw San Antonio and its eastside as places of hope and socioeconomic opportunity.[17] Many of these persons were related to the Ball family, who lived either in Bexar County or in farm communities in adjacent counties to the south and east of Seguin at Sweet Home Road (in Seguin) and Zion Hill Road (in Guadeloupe). Their families include such surnames as Ball, Dibrell, George, Hysaw, McIntyre, Reddix, Ruffin, Scott, Sheffield, Smiths, Sutton, Williams, and Woolridge, to name a few—families with South Texas roots that trace back to antebellum Texas in five colonies around Zion Hill Baptist Church in Guadalupe.[18] The Balls were one of several black families that migrated to San Antonio from the proximity of their ancestral homes in rural Texas during the postwar period. This group found hope with the passage of the 1964 Civil Rights Act, 1965 Voter Rights Act, and civil rights activism of local leaders like Antioch Missionary Baptist Church's Reverend Claude Black.[19]

One civil rights law often overlooked by scholars is the 1968 Civil Rights Act. This law not only provided enforcement for previous civil rights bills but also included the National Fair Housing Act of 1968 under Title VII. According to historian Lawrence de Graaf, the passage of this law did two things: first, it "prohibited most forms of discrimination based on race, color, religion, or national origin in the sale, rental, or financing of housing"; and second, it "made [FHA] . . . insurance more accessible to African American home buyers and lower-income families, and Congress subsidized home purchases and renting by low-and moderate-income families under sections 235 and 236 programs of the 1968 Housing Act."[20]

In western metropolises, this law sparked black suburb growth on the outskirts of Anaheim, Arlington, Los Angeles, Phoenix, and San Jose, to name a few places.[21] This occurred just as these suburban metropolises were about to explode in growth and in population—so much that the most dynamic growth in postwar American cities has been in the West, a place that horizontally stretches from Hawaii to states vertically aligned on the 98th meridian, including Texas and Oklahoma.[22] Today, Texas has three cities in the top ten most-populated U.S. cities: Houston, San Antonio, and Dallas, with the Alamo City ranked in seventh place at 1,327,407—91,280 of which are African Americans, at 7 percent of San Antonio's population.[23] As these cities have grown, so have the populations in their centers and suburbs. However, what makes the Alamo City different from other large western cities is that its suburbanization exploded in growth only after 1980, twenty to thirty years after other similar-sized cities began expanding their city limits.[24]

San Antonio's suburban explosion was set in motion after HemisFair '68 by its city planners, Good Government League officials, and business elites such as local chamber of commerce chairman B. J. "Red" McCombs and banker Tom Frost. According to most city plans, San Antonio's urban growth was viewed as a priority in the 1970s. Based on the plans, urban growth was to occur in every sector of the city, via the revitalization and development of resources that made each section unique, such as the most obvious, revitalizing the Alamo-Riverwalk area along the San Antonio River. Linking this growth to the city's outskirts was supposed to be a combination of freeway and rapid transit systems in a pattern resembling the spokes of a wheel.[25] During the 1970s, recession and stagflation ended most meaningful talk of revitalizing the eastside ghetto and westside and southside barrios. After the 1970s, most urban redevelopment meant revitalizing areas linked to the Alamo-Riverwalk area for nostalgic tourism and for professionals willing and able to purchase inflated-priced condos and lofts in adjacent communities.

In the eastside, because African Americans have the smallest ethnic population by percentage, politically and economically they are arguably "the most ignored group in San Antonio."[26] Since 1980 this had led to stagnation in east San Antonio's growth. Stagnated growth has also impoverished the westside and southside, despite the fact that (since 1980) Latin Americans have been growing in importance as the city's majority population at a range between 52 and 63 percent from 1980 to the present.[27] Within this group, Mexican Americans are 93 percent of the population. Still, when we combine the urban development efforts of every region that is not the northside, what becomes glaringly obvious is that no group in the west, south, or east has formed sustained revitalization efforts to build on community resources and promote community growth on a scale compared to that in the city's north and downtown areas. As a result, poorly planned eastside revitalization schemes like the development, construction, and opening of the AT&T Center (in 2002) have failed to spark sustained, local socioeconomic growth. The Center instead appears primarily to benefit its owners' economic bottom line, Riverwalk tourism, and restaurants and shops from the North Star Mall (near the airport) to malls in the northwest. According to San Antonio social marketer and freedom rights activist Tommy Calvert, San Antonio is about thirty years behind most major cities in its inner city redevelopment.[28] Moreover, eastside business growth has been stunted, in part because of a lack of a local planning group.[29] Ultimately, this has resulted in east San Antonians being politically and economically ignored, socially isolated, pejoratively stereotyped by the media, and victims of de facto racial discrimination.

Since 1990, the eastside's reputation for crime, "blighted" properties, subpar public education, and gang activity has pushed promising black youth away from the area. Many are leaving as college students who bypass city community colleges like the eastside's St. Phillips College for higher education in the suburbs, or they are leaving what many people perceive as a decaying section of the city for living-wage opportunities and cities with vibrant reputations like Houston and Dallas. For former eastside residents like James Alfred Ball, many San Antonians live in the eastside because they either lack the financial means to leave, want to improve it like Tommy Calvert, or are comfortable there (even as they may have the means to move out).[30]

AFRICAN AMERICAN COMMUNITIES ON THE OUTSKIRTS OF ALAMO CITY

Since the 1980s, blacks have moved into San Antonio suburbs. In this period, four categories of African Americans have led this movement: retired veterans, professionals, active military personnel, and students. I explore this topic through the lives of black suburbanites related to or closely affiliated with the Ball family, who, as previously mentioned, are one of greater San Antonio's indigenous African American families, with a population of more than 1,300.[31]

Black veterans have been settling in inexpensive suburban subdivisions near medical centers since 1980. As a group, veterans like Vernon "Honey" Ruffin have been making east San Antonio their home since Brooke Army Medical Center opened its doors to black veterans in the 1950s during the Korean War. This medical center garrisoned at Fort Sam Houston in the northern section of east San Antonio is a teaching hospital. Today it is called the San Antonio Military Medical Center and is part of the University of Texas Health Science Center at San Antonio (headquartered in the northwest) and U.S. Army Medical Command. The presence of this medical industrial complex was immense in promoting the population growth of San Antonio's city center from 1945 to 1980, and since 1980 in reshaping the way San Antonio is developing toward the seven counties that surround Bexar County.[32]

An African American pioneer in this category is James Shelby Ball, a Vietnam veteran who contacted Agent Orange while in the field of duty as a medic. Unlike most black San Antonians drawn to the city through the military, similar to Honey, James Shelby is a native to the greater San Antonio region. Previous knowledge of the city was important in his pioneering role in the 1970s as a person of African descent moving into the formally prohibitive northwest section of the city. However, the part of the northwest that he and his family moved to was Leon Valley, which was semirural and remote prior to 1997. This was the time when freeway, subdivision, and strip mall construction exploded toward the outskirts of northwest and north-central San Antonio. What made this particular area appealing for James Shelby was its proximity to his employer, the University of Texas Health Science Center at San Antonio, and its relative calm. Since the suburbanization of the northwest, James Shelby has had to trade relative calm for rising home values and shopping convenience. In the larger scheme, James Shelby's family were pioneers because there was only one census tract where blacks had a minute presence, near Fredericksburg Road and Huebner Road, at 3 percent.[33] Since 1990, every region between I-410, Loop 1604, I-10, and State Highway 151—which makes up the western half of northwest San Antonio, or the far northwest (i.e., Leon Valley, Braun Station, and Parkwood)—had a similarly small African American population. In this area, the one community where the black population rose above 10 percent in 2000 and 2010 was between Fredericksburg Road and the I-10.[34]

The core institution uniting the black San Antonio suburban community is the black church. In the northwest, St. John's Baptist Church is one of several black churches that unite African Americans rooted in the southern evangelical tradition. Although the church is not a megachurch—which are growing throughout north San Antonio—like the megachurches St. John's attracts church membership from all over the city.[35] Not only does this make it convenient for members to get the spirit and a sense of cultural community, but suburban churches are also great places for former friends, neighbors, and colleagues to reunite. According to church member Rae Lowry, most St. John members live in the proximity of the far northwest, with a minority in east and northeast San Antonio.[36]

Representing the second category of black suburbanite, Rae Lowry is a retired AT&T customer service manager who has lived in northeast San Antonio since the start of its suburban development in the mid-1990s. In the 2000s she was drawn to St. John's after being invited to attend by a friend who lived in Leon Valley. Prior to being a member of St. John's, Rae was a member of the "silk stocking" black church (or black middle class) Antioch Missionary Baptist Church in east San Antonio. Antioch church members include a who's who of black San Antonio's small middle to upper middle classes, such as San Antonio Spurs players, businessmen, judges, and lawyers. Since the 1980s, as these affluent African Americans have increasingly migrated to the northern suburbs, so too have their weekly Sunday treks into the eastside increased so that they could hear the word, tithe, and socialize with people they culturally identify with at black institutions steeped in local tradition, like Antioch Missionary Baptist Church, Mount Zion First Baptist Church, and New Light Baptist Church. Lowry found St. Johns (a lower black middle-class church) to be more comfortable, because of friendships and the practical message "that you study and work in the church."[37]

Like Honey Ruffin and his cousin James Shelby Ball, Rae Lowry is a native of rural south-central Texas. She was born in 1950 and raised in the farm town of La Vernia, eighteen miles east of the nearest I-410 exit in east San Antonio. In this town, her family grew close to the Ruffins and Balls, who lived in nearby farm communities. Unlike Honey, who still lives in the eastside, and James Shelby, who still lives in the northwest, Rae lives in northeast San Antonio (east of I-35). This is the area where most black San Antonio suburbanites live, in small yet rapidly growing cities which in 2010 ranged between 3,000 and 32,000 people: Cibolo, Converse, Garden

Ridge, Kirby, Live Oak, Schertz, Selma, Universal City, and Windcrest.[38] These cities rest between I-410, Loop 1604, and I-35, and I-10 heading toward New Braunfels and Seguin. In this region, the 2000 and 2010 black suburban populations ranged between 7 and 21 percent. This is larger than black San Antonio's population average at 7 percent. Most African American suburbanites like Rae treat the communities they reside in as traditional bedroom communities, with the caveat that when they want to be entertained they socialize in nearby east San Antonio, to "hear [their] kind of music. And meet [their] kind of people."[39]

Rae's comfort in living in a predominantly white suburb while staying connected with black people in the eastside is rooted in her upbringing in south-central Texas. In La Vernia her tentative comfort in being around persons of European descent began during adolescence, when she played with white children. However, she was also raised to understand the limits of racial integration. For example, when black and white children played together, they always played at the black children's homes and never in the white children's homes. As this was occurring, by the early 1960s, racial barriers in greater San Antonio noticeably began to collapse. In towns like La Vernia, African Americans started accessing white-owned restaurants. In education, schools like Marion High School (in nearby Marion), where Herb Ruffin attended in 1963–65, and La Vernia High School, where Rae attended in 1964–68, systematically opened their doors to students of color under federal orders to desegregate.[40]

Despite living under Jim Crow conditions, because the Lowrys and Ruffins represented a small percentage of African Americans that posed no threat to the white socioeconomic order in towns like Marion, La Vernia, and New Berlin, they were essentially tolerated and respected for their ethic of hard work as independent farmers. The end result for many local blacks who were part of Rae's and Herb's generation was that they became tentatively comfortable living in a culturally diverse world that included Latin Americans. In 1968, after graduating from high school, Rae and her twin sister moved to east San Antonio—which was common for postwar black youths from nearby farm towns to do. Many of these youths with Southwest racial sensibilities were lured to the Alamo City by the potential of living comfortable urban and industrial lives devoid of farming, rattlesnakes, and dependency on capitalistic markets and Mother Nature.[41]

In 1975, Rae left the Alamo City for socioeconomic opportunities with Southwestern Bell Corporation (later SBC Communications, and AT&T)

in nearby Austin. Twenty years later (1995), a management promotion within AT&T's customer service repair center brought Rae back to San Antonio. During her absence the city had grown phenomenally, expanding its borders from within the I-410 to Loop 1604 through annexation and rapid urban development. Rae and her former husband especially saw most of this activity occurring in the northwest and north-central areas. For Rae's family, their first question moving back was what community had the combination of a location near their jobs, family, and grocery stores like H.E.B. and good resale value. Keep in mind, implicated in the resale question was living in a safe and clean neighborhood with good amenities, good public schools, and where good equity could be amassed. Their second question concerned not disconnecting themselves from the eastside black San Antonio community. In this Rae and her husband compromised by not living in a gated community—which a few of her black friends and a huge population of north San Antonians live in—so that their family could drive in and out without feeling harassed by guards at their subdivision. After making their decision, Rae's family paid Swientek Construction to build their home, making them the second household to live on their block. As for achieving the American Dream, seventeen years after moving into the northwest Rae says she feels that she has, and she gave these reasons: "I own my own home. Have money in the bank. I live comfortably. . . . I am not poverty stricken. I have a car. I'm OK, [because] my dream could be what I want it to be."[42]

Rae's decision on where to settle was one of many different paths African Americans took in post–civil rights era San Antonio (since 1970). Crucial to this development were the civil rights victories that opened the doors for blacks in most communities with the passage of the 1968 Fair Housing Act. Over a generation later, time has been another crucial factor in changing white attitudes about living in communities with people of color. According to the people interviewed for this article, relatively more white people don't mind living next to middle-class people of color. Rae feels that her white neighbors "don't have the Jim Crow feel." She also mentioned that "a new day has appeared to emerge in San Antonio, in terms of growth. Black suburbanites are in the middle of this growth."[43] As glowing as that forecast on local race relations was, the same could not be said for the majority of whites living next to working-class people of color in west, south, and east San Antonio.

Despite the change in housing laws, time, and feel, Rae's analysis is challenged by the fact that, since 2000, older suburban communities like

Alamo Heights, Terrell Hills, and communities within the Golden Triangle have barely begun to diversify racially. African American presence in these areas ranges from less than 1 to 3 percent; "diverse community" has come to mean predominantly white with a few Mexican American neighbors—a pattern that developed after 1995 in most suburban communities outside of northeast San Antonio. As for whether racial discrimination in housing was occurring to produce such a peculiar suburban-urban pattern steering blacks to the eastern half of San Antonio, African American suburbanites interviewed were unsure.

What everyone is sure of is that black suburban sprawl from east San Antonio is gradually penetrating communities above Loop 1604. This is an unprecedented movement, because communities there have been high-property-value havens for white suburbanites in north San Antonio. Following already described trends, black suburbanites are moving into upper northeast San Antonio. The family of Jesse Ball, cousin of James Shelby and Honey Ruffin, moved to the upper northeast recently. Like James and Honey, Jesse is a native to the region, raised in the oldest branch of the Ball family south of Seguin. Like many Alamo City black suburbanites, Jesse's family is more representative of active military personnel settling and retiring in San Antonio. For many active duty personnel, their first exposure to San Antonio was during their visit to a medical center or facility of the armed forces, in which the city has a strong military presence: northwest at Camp Bullis, upper northeast at Randolph AFB, eastside at Fort Sam Houston, southeast at Brooks City Base, and southwest at Lackland AFB and Kelly AFB (until 2001, which is now Kelly Field annexed to Lackland AFB).[44] In the 1970s, before the suburbanization of the northeast, active military personnel and retired veterans initially moved into suburbanized communities near these bases. The first African American pioneers to move into the western portion of San Antonio, beyond I-410, were active duty personnel and retirees affiliated with Lackland AFB. According to the census, from I-410 to Loop 1604 and I-90 up to State Highway 16/Bandera Road, the black presence has run consistently in most communities at around 14 percent.[45]

Like most black veterans exposed to the city, Jesse first lived in the eastside. Since the late 1990s that pattern has been changing as more African American military personnel and retired veterans are moving into the rapidly suburbanizing northeast. Jesse's primary reason for moving to the upper northeast was to own good property in a nice and safe community near family and the San Antonio Military Medical Center. What stood out in this discussion was Jesse's emphasis on safety and the virtues of gated communities, even though his family lives in a nongated community.[46] According to reporter Haya El Nasser, Jesse's (and Rae Lowry's) emphasis on safety and high regard for gated communities, though they were reluctant to totally embrace them in 2002, runs consistent with most affluent African American attitudes. At the time, "the popularity of gated communities [was] on the rise nationwide, according to developers and housing experts. In a nation still confronting post-9/11 jitters, living behind walls and knowing your neighbors create[d] a safety zone for many. Security [was] also a top concern for baby boomers as they head[ed] toward retirement."[47]

According to the census bureau's 2001 American Housing Survey, gated developments were more prevalent in the Sun Belt metro areas such as Dallas, Houston, and Los Angeles than in older urban areas in the Midwest and Northeast. Despite these developments, the final analysis of Nasser's report found that blacks were "less likely to live in gated communities than whites and Hispanics" because it simply did not feel right after being excluded from such communities for so long.[48] Many of the black baby boomers who experienced Jim Crow, and who make up the bulk of the current movement into the suburbs, are doing what demographers and sociologists are urging other Americans to do—not isolate themselves from a diverse range of people behind walled communities. Black San Antonians with familial and friendship ties to other parts of the city appear to be making conscious decisions about where they live and who they relate with.

Another place where the gated phenomenon is rampant is northwest and north-central San Antonio. This area is the original home of San Antonio's business elite, which are overwhelmingly white. Even though the assessment of historian Frank Char Miller is well noted—that the same lily-white demographics exist in these areas as during the early twentieth century—a fourth development in black suburban San Antonio is gradually emerging, with the rise of the black middle class that either does not have roots in greater San Antonio or is younger, has means, and has a post–civil rights era outlook that allows them to move into formerly racially restrictive areas. This outlook is an incredible about-face for people from a region which, before 2000, represented the lowest percentage of middle-class professionals in San Antonio by race and region. Furthermore, they come from the weakest political group in the Alamo City, including the far more numerous Latino population, which has produced three mayors of Mexican descent since 1980 in Democrats Henry G. Cisneros (1981–89), Edward D. Garza (2001–5), and Julian Castro (2009–14). Ultimately, the eastside's underdevelopment and the gradual loss of hope has encouraged many black middle-class people and promising youth to move to communities that are perceived to offer them the best route to socioeconomic freedom.[49]

Prior to rapid suburbanization, St. Philips College, a two-year college, was the main institution of higher education that provided eastside African Americans the keys to socioeconomic mobility. Since 1995 that has noticeably declined, not because fewer blacks are attending college but because more are attending predominantly white four-year colleges/universities and are, in particular, leaving the eastside for north San Antonio to go to school. This pattern reflects what has been happening in San Antonio since the 1980s, with explosive suburbanization in the north and rapid expansion of the University of Texas at San Antonio's (UTSA) main campus in northwest San Antonio.[50] Today, many African American college graduates are not going back to the eastside, except to visit their families and friends or to work noncareer jobs and live in their families' homes until their next paths toward upward mobility unfold. The rise of this black flight population mirrors the growth of the region. According to the *San Antonio Express-News* there has been a large flight of African American college and K-12 students to the northern suburbs outside of I-410—if they stay in the city.[51] Otherwise, as previously stated, most black San Antonio professionals and prospects often unknowingly repeat the path of their predecessors and leave the Alamo City for more cosmopolitan-feeling cities like Houston, Dallas, and, increasingly, Austin.[52]

Less common are young African Americans like Honey Ruffin's daughter, Jozette, who graduated with advanced degrees in academic counseling and is determined to make a difference in the eastside public school system. In this regard, she follows in her mother's footsteps. Her mother is retired math teacher Delores Ruffin. Prior to Jozette's father's passing in 2017, the home she lived in was across the street from her father, who owned enough properties to move his family out of the aging eastside, though the family would not leave when he lived. Similar to eastside political activist Tommy Calvert, they are one of many black families who are the heart of eastside and central city San Antonio. The same cannot be said of most young black professionals, who, as new college graduates, are either retained in the

Alamo City in jobs that represent economic mobility after graduating from UTSA, Trinity College, or (more likely) one of the six Alamo community colleges or were recruited to work in a broad plethora of entry-level, civil servant, and middle-management positions at the armed forces, AT&T, City of San Antonio, or United Services Automobile Association. According to Rae Lowry, most of these professionals are finding homes either in suburbs or in revitalized condos and lofts around the Alamo-Riverwalk area.[53]

Before now, African American youth saw San Antonio as a retirement city. This was the view of professionals like Rae, veterans like Jesse Ball, and retired craftsmen and military veterans like Rae's current significant other, Jesse's cousin Herb Ruffin. As of this writing Herb has not moved to San Antonio, but he has been contemplating the move since the dot-com bubble burst in 2000. Herb is originally from a farm town east of San Antonio called New Berlin. In 1967 he left the greater San Antonio area to serve in the air force. He volunteered to avoid being recruited into the army and probably serving on the front lines in Vietnam, around the time that the war was escalating and black soldiers were being drafted to fill military manpower needs under Project 100,000.[54] Instead, Herb briefly worked in Okinawa, Japan, as an airplane mechanic and was commissioned to serve at Beale Air Force Base in Northern California. In the early 1970s after his release from duty, he replanted his roots in the Bay Area. The city he has always been associated with since his release is East Palo Alto—a working-poor African American community populated by many south-central Texans related to the Balls and Ruffins.[55]

Today, even a working-poor community in Silicon Valley such as East Palo Alto is expensive to live in, with the average four-bedroom house costing about $50,000 more than most similar-sized homes in the affluent Golden Triangle and at least three times more than most four-bedroom homes in San Antonio.[56] This along with gentrification and hyperinflation are factors pulling Herb back to the much less expensive San Antonio area. The area he wants to move to is along the I-35 corridor in the northeast, which is central to his family and significant other and is where the next great suburban boom is taking place—linking San Antonio to New Braunsfeld and Austin. Whether this latest boom will be a continuous stream of suburban communities similar to the outskirts of other Southwest cities remains to be seen. If Californian migrants and investors have their way, north San Antonio will become another California, in which the time is now to purchase a home before the cost explodes from $100,000 for a new four-bedroom single-family home to over $400,000, which is what happened in California from 1978 to the early 1990s.[57]

During the 1990s, many Californian residents left the San Francisco Bay area and Southern California for better housing and economic opportunities. Many of these people were speculators, or residents who used their homes as a commodity to cash out and use the equity to pan for another metropolitan market where they could live their "American dreams" as suburban homeowners. Although this gold rush mentality has taken many Californians to affordable outskirt communities such as the upper Contra Costa Valley and San Bernardino County, most have become crucial in the growth, hyperinflation, and overpopulation of accidental suburban metropolises like Phoenix and Las Vegas. Since 2005, this Californian movement into Phoenix and Las Vegas exurbs has slowed down, only to find new housing markets in the interior Southwest cities of San Antonio, Austin, and Oklahoma City. According to Trinity University librarians Meredith Elsik and Jessica Tamayo, in San Antonio the influx of new migrants to the north of the city is dramatically changing the way the city is developing.[58] As a result, San Antonio's suburbanization process has accelerated, splitting the city between the modernizing north, the stagnated west, south, and east, and the heritage-developing Alamo-Riverwalk area.

EPILOGUE: CLASS DIVIDE IN BLACK SAN ANTONIO?

Even though San Antonio is trifurcating, remarkably this process has only minimally split black San Antonio. Why has this socioeconomic split that is creating multiple Alamo Cities not yet occurred in its African American communities? For one thing, black San Antonio has a small yet growing middle class. For another, most African Americans are part of what can be referred to as the "new black working class." Finally, most black suburbanites are older (baby boomers), have previous affiliation with San Antonio, or are coming back because of their family and friends and the Alamo City's inexpensive cost of living.

Concerning the first factor, although the black professional class has been growing since the 1970s, because San Antonio has had a hard time retaining African American professionals up until recently this class has been minute. Although most black middle-class people are reflected in these three factors, the biggest factor that separates them from the white middle class is the acquisition of inherent wealth.[59] The net worth of most African Americans is based on paychecks and not intergenerational wealth, which since the World War II has been closely tied to homeownership in high-valued communities such as Alamo Heights and those in northwest San Antonio. For most blacks in the middle-income bracket, a bad break or economic recession as in 2008 could easily drop them into the ranks of the foreclosed and working poor. As for occupation, most black middle-class San Antonians work as managers, professionals, and sales workers in white-collar occupations. Before African Americans were obtaining these jobs, occupations such as clerk, porter, and unionized factory worker—like Honey Ruffin and his brother Willie "Hooley" Ruffin—were seen as middle-class professions. Since the 1980s that has not been the case in the face of eastside manufacturing employers, such as Jordan Ford and Pearl Brewing Company, systematically folding around the time real wages declined and the average person's cost of living rose.[60] In the 1990s, a similar pattern has been occurring for white-collar employees, leaving one to ask, in this new millennium, "What is San Antonio's black middle class"?

In 2000, most suburbanites within San Antonio's black middle class lived on median family incomes ranging from $35,000 to $40,000. Many of these people were retired and lived on fixed incomes, which was $6,000–$10,000 more than black San Antonio's median income of $29,598.[61] With that said, most African American suburbanites are solidly part of the black lower middle class. They function, along with the working poor, as a "new black working class," as noted above. In the 2000s, many people in the upper income bracket of this new black working class live lives closer to their working-poor counterparts than to the white middle class that lives on median incomes above $50,000.[62] Even as this occurs, both the black lower middle class and most of the black working poor still believe that they are middle-class people, though the core marker of that status—to live comfortably, exceeding one's needs, is limited. The bottom line here is that San Antonio's black lower middle class and working poor strongly appear to identify with one another, despite living in different communities.

Essential to this identification is that former eastside residents maintain strong ties to the people in their former community. Through steady affiliation with people within the eastside and reticence to buy into the predominately white gated community phenomenon, most San Antonio black suburbanites have developed effective resistance tactics to potential social isolation on the outskirts of the Alamo City. Moreover, what appears to be "ring suburban" development, or black community extension from the eastside to the northeast, is not actually so.[63] Black San Antonio suburbanization is unique in its formation. Many African American suburbanites who left the city center for opportunities beyond San Antonio are moving back into newly developing suburban communities in north San Antonio. This is not merely a movement from the eastside to the northeast but a mixture between that and an older generation coming back to the city, often seeing themselves as living within city limits (even beyond I-410) while not distancing themselves from former acquaintances. With that said, like Rae Lowry and Jesse Ball, many black suburbanites are consciously purchasing homes in locations central to their needs, which are near US 281, I-410, and I-35. These freeways quickly take black suburbanites to the eastside, black churches, markets that sell Louisiana sausages, restaurants like Tommy's Soul Food, and traditional socializing spots like Lincoln Park and the Little Fish Factory. In the final analysis, black suburbanization in San Antonio appears to be deeper than the process we normally think of.[64] For most African Americans native to south-central Texas, black suburbanization in the Alamo City has come to represent lives that have come full circle from adolescence to adulthood in close proximity to their American ancestral homelands.

NOTES

1. Alwyn Barr, *Black Texans: A History of African Americans* (Norman: Oklahoma University Press, 1996); Larry P. Knight, "Defending the Unnecessary: Slavery in San Antonio in the 1850s," in Bruce A. Glasrud and Cary D. Wintz, eds., *African Americans in South Texas History* (College Station: Texas A&M University Press, 2011), 29–46; Robert C. Fink, "Semi-professional African American Baseball in Texas before the Great Depression," in Bruce Glasrud and James M. Smallwood, eds., *The African American Experience in Texas* (Lubbock: Texas Tech Press, 2007), 218–30; Kenneth Mason, *African Americans and Race Relations in San Antonio, Texas, 1867–1937* (New York: Routledge, 1998); Jerelyne Castleberry Williams, *The Brackenridge Colored School: A Legacy of Empowerment through Agency and Cultural Capital inside an African American Community* (Bloomington, IN: Authorhouse, 2006); Judith Kaaz Doyle, "Maury Maverick and Racial Politics in San Antonio, Texas, 1938–1941," in Glasrud and Wintz, *African Americans in South Texas History*, 206–41; Robert A. Goldberg, "Racial Change on the Southern Periphery: The Case of San Antonio, Texas, 1960–1965," in Glasrud and Wintz, *African Americans in South Texas History*, 280–312.

2. Albert S. Broussard, *Black San Francisco: The Struggle for Racial Equality in the West, 1900–1954* (Lawrence: University Press of Kansas, 1993); Sheryll Cashin, *The Failures of Integration: How Race and Class Are Undermining the American Dream* (New York: PublicAffairs, 2005); Lawrence B. De Graaf, "African American Suburbanization in California, 1960 through 1990," in Lawrence B. De Graaf et al., *Seeking El Dorado: African Americans in California* (Seattle: University of Washington Press, 2001), 405–49; Emory J. Tolbert and Lawrence B. de Graaf, "'The Unseen Minority': Blacks in Orange County," *Journal of Orange County Studies* 3, no. 4 (Fall 1989/Spring 1990): 54–61; Karyn R. Lacy, *Blue-Chip Black: Race, Class, and Status in the New Black Middle Class* (Berkeley: University of California Press, 2007); Thomas J. Sugrue, *The Origins of the Urban Crisis: Race and Inequality in Postwar Detroit* (Princeton, NJ: Princeton University Press, 1996); *Sweet Land of Liberty: The Forgotten Struggle for Civil Rights in the North* (New York: Random House, 2008); Quintard Taylor, *In Search of the Racial Frontier: African Americans in the West, 1528–1990* (New York: W. W. Norton, 1998); Quintard Taylor, *The Forging of a Black Community: Seattle's Central District, from 1870 through the Civil Rights Era* (Seattle: University of Washington Press, 1994); Joe William Trotter, *Black Milwaukee: The Making of an Industrial Proletariat, 1915–1945* (Urbana: University of Illinois Press, 1985); Joe William Trotter, ed., *The Great Migration in Historical Perspective: New Dimensions of Race, Class, and Gender* (Bloomington: Indiana University Press, 1991); Matthew C. Whitaker, *Race Work: The Rise of Civil Rights in the Urban West* (Lincoln: University of Nebraska Press, 2007); Andrew Wiese, *Places of Their Own: African American Suburbanization in the Twentieth Century* (Chicago: University of Chicago Press, 2004).

3. The trifurcation concept is inspired by interviews by the author with Trinity University librarians Meredith Elsik and Jessica Tamayo and University of Texas at San Antonio managing director of university communications Lety Laurel in July 2012, San Antonio. For these well-informed natives, the Alamo City has split between people living within the 410 beltway and those outside of it. Moreover, the only place that suburbanites are comfortable going to is the Alamo-Riverwalk area.

My nicknaming San Antonio as the "Alamo City" is rooted in what it officially sold itself to the public as, prior to 2012, before the Department of Defense's Base Realignment and Closure (BRAC) Commission invested billions of dollars into the city for the redevelopment of its military installations and communities adjacent to those bases and medical facilities. From 2012 to 2015, this political economic development, designed to boost San Antonio's economy by protecting its military industrial complex, has resulted in the city officially renaming itself "Military City USA." For more on this recent development, see San Antonio Convention and Visitors Bureau, "Military USA," http://visitsanantonio.com/MILITARY; Mike Gallagher, "S.A. Must Preserve 'Military City, USA,'" in *My San Antonio*, July 9, 2015, www.mysanantonio.com/opinion /commentary/article/S-A-must-preserve-Military-City-USA-6376357.php; and Sig Christenson, "S.A., Texas Could Gain in Wake of Troop Cuts, BRAC," in *San Antonio Express News*, March 3, 2014, www.expressnews.com/news/local/military/article/S-A -Texas-could-gain-in-wake-of-troop-cuts-BRAC-5285609.php.

4. "Old Road Maps of San Antonio," *TexasFreeway.com*, www.texasfreeway.com /sanantonio/historic/road_maps/sanantonio_road_maps.shtml; "San Antonio Free-way System: Interstate 410 (John B. Connally Loop)," *TexasHighwayMan.com*, www .texashighwayman.com/i410.shtml; "San Antonio Freeway System: State Loop 1604 (Charles W. Anderson Loop)," *TexasHighwayMan.com*, www.texashighwayman.com /lp1604.shtml.

5. See William H. Frey, "Melting Pot Suburbs: A Study of Suburban Diversity," in Bruce Katz and Robert E. Lang, eds., *Redefining Urban and Suburban America: Evi-dence from Census 2000* (Washington, DC: Brookings Institution Press, 2003), 175–78. Census and map data and conclusions throughout this chapter are based on the sub-scription website *Social Explorer* (www.socialexplorer.com) and U.S. Census Bureau website on the Decennial Census of Population and Housing, 1950–2010, www.census .gov/programs-surveys/decennial-census/decade.1950.html.

6. *Social Explorer*, s.v. U.S. Demographic Maps, Census 2010.

7. Based on interviews by the author with Meredith Elsik, Jessica Tamayo, and Lety Laurel, July 2012. Other interviews by the author that contribute to this chapter: James Shelby Ball, Jesse Ball, and family, July 2012; Herb Ruffin I, October 2011, July 2012, November, 2012; Rae Lowry, November 2012; Honey Ruffin, Delores Ruffin, and Jozette Ruffin, July 2009, July 2012.

8. City of San Antonio, "City of San Antonio Growth by Annexations Every Tenth Year," www.sanantonio.gov/Portals/0/files/GIS/Maps/Annexation_8.5x11.pdf (accessed February 24, 2019.

9. Miller quoted in Nicole Foy, "Detached but Engaged—Despite Boundaries, Suburbs Maintain Stake in San Antonio," *San Antonio Express-News*, September 5, 1998.

10. City of San Antonio Department of Planning, "Annexation, Public Facilities and Services," in "Draft: Land Use Plan—City of San Antonio" (October 1982), 47; and City of San Antonio, "East Side Planning District: Background Information" (1973) (Coates Library, Trinity University, San Antonio), 3.

11. *Social Explorer*, s.v. Tables, Census 1980–2010, tracts 1209.1, 1211.1, 1212.1, 1212.2, 1214, 1215, 1218, 1315, 1316.2; s.v. Tables, Census 1990–2010, tracts 1210, 1211.04, 1218.05, 1218.01, 1209.1, 1212.01, 1212.02, 1218.02–.04, 1215.01–.07, 1216.03, 1216.04, 1214.01–.04, 1315.01–.02, 1316.03–.07. See also *Censusviewer.com*, "Con-verse, Texas Population," http://censusviewer.com/city/TX/Converse, and, on the

same website, population tables for the towns of Kirby, Live Oak, Universal City, and Windcrest.

12. Rae Lowry, Meredith Elsik, Jessica Tamayo, and Lety Laurel interviews. See also Callie Enlow, "Left Behind: Why People Leave San Antonio," *Rivard Report*, April 17, 2012, http://therivardreport.com/left-behind-why-people-leave-san-antonio.

13. Office of Preservation, *City of San Antonio*, "King William," www.sanantonio .gov/historic/Districts/King_William.aspx; Christopher Long, "The King William Historic District" *Texas State Historical Association*, www.tshaonline.org/handbook /online/articles/ghk01.

14. James Shelby Ball, Jesse Ball, Honey Ruffin, Rae Lowry, Meredith Elsik, Jessica Tamayo, and Lety Laurel interviews. See also John Tedesco, Elaine Ayala, and Brian Chasnoff, "As S.A. Grows, Folks Go North and West," *San Antonio Express-News*, March 6, 2011.

15. *Social Explorer*, s.v. Tables, Census 1980–2010; and s.v. U.S. Demographic Maps, Census 1980–2010 (variables: "Race," "Black," and "Hispanic or Latino" by percentage).

16. Honey Ruffin and James Alfred Ball interviews. See also Goldberg, "Racial Change on the Southern Periphery"; and *SNAP* (Black San Antonio weekly, 1955–66), in Lewis F. Fisher, San Antonio Black History Collection, 1873, 1923–1996, University of Texas at San Antonio Special Archives, UTSA Library, San Antonio.

17. Honey Ruffin, Delores Ruffin, James Shelby Ball, Jesse Ball, James Alfred Ball, Rae Lowry, Alfred Smith, Ron Hall, and Earl Reddix interviews.

18. Based on annual Ball family reunion pamphlets (2009, 2010, 2012); Ball family historian, James Alfred Ball; Ruffin family historian, Honey Ruffin; and guided tour, images of gravestones, and interview given to author by Ball family member Ron Hall, July 2012.

19. "Interview with Rev. Claude Black, 04–11–2006," UTSA Library, Digital Collections, http://digital.utsa.edu/cdm/singleitem/collection/p15125coll4/id/2014/rec/1; interview given to author by San Antonio NAACP president Oliver Hill, November 2015; Goldberg, "Racial Change on the Southern Periphery," 280–312; Claude and Zer-Nona Black Papers, 1890–2009, Trinity University Archives and Special Collections, Coates Library, Trinity University, San Antonio.

20. De Graaf, "African American Suburbanization," 415; Civil Rights Act of 1968, Title VIII (Fair Housing Act of 1968). 42 U.S.C. §3601.

21. For more on modern urban and postsuburban development in the West, see De Graaf, "African American Suburbanization," 343–76; Allan A. Saxe, *Politics of Arlington, Texas: An Era of Continuity and Growth* (New York: Eakins Press, 2001); Josh Sides, *L.A. City Limits: African American Los Angeles from the Great Depression to the Present* (Berkeley: University of California Press, 2006); Whitaker, *Race Work*; and Herbert G. Ruffin II, "The Search for Interstitial Space: San Jose and Its Great Black Migration, 1941–1968," in Ingrid Banks and Clyde Woods, eds., *Black California Dreamin': The Crises of California's African-American Communities* (Santa Barbara, CA: UCSB Center for Black Studies Research), 19–56.

22. See Richard White, "Western History," in Eric Foner, *The New American History* (Philadelphia: Temple University Press, 1997), 203–30; and Taylor, *In Search of the Racial Frontier*, 17–23.

23. "Top 20 Cities: Highest Ranking Cities, 1790 to 2010—San Antonio," *U.S. Census Bureau*, www.census.gov/dataviz/visualizations/007. San Antonio has been

a top-twenty U.S. city since 1960: seventeenth in 1960, fifteenth in 1970, eleventh in 1980, tenth in 1990, ninth in 2000, and seventh in 2010.

24. Robert E. Lang and Patrick A. Simmons, "'Boomburbs': The Emergence of Large, Fast-Growing Suburban Cities in the United States," Fannie Mae Foundation Census Note 6, June 2001, 1. For additional information on the acceleration of suburbanization, see Robert E. Lang and Jennifer Lefurgy, *Boomburbs: The Rise of America's Accidental Cities* (Washington, DC: Brookings Institution Press, 2009); Joel Garreau, *Edge City: Life on the New Frontier* (New York: Anchor Books, 1992); Andres Duany, Elizabeth Plater-Zyberk, and Jeff Speck, *Suburban Nation: The Rise of Sprawl and the Decline of the American Dream* (New York: North Point Press, 2010); and Wiese, *Places of Their Own*, 255–92.

25. Alamo Area Council of Governments, "Regional Development: Alternative Growth Patterns" (1969), 4, 9–24; San Antonio City, "State of the City: '72 San Antonio," both in Coates Library, Trinity University.

26. Jesse Ball interview.

27. *Social Explorer*, s.v. Tables, Census 1980, R10390393; s.v. Census 1990, R10390398; s.v. Census 2000, R10390401); and s.v. Census 2010, R10390403.

28. Tommy Calvert Jr., "Look to the Eastside for San Antonio's Revitalization," *Rivard Report*, August 23, 2012, http://therivardreport.com/look-to-the-eastside-for-san-antonios-revitalization.

29. Dennis Branch, "The Lack of Economic Development on the Eastside of San Antonio," Master's thesis, Trinity University, 1983; Bob Wise, "Building San Antonio; It Takes a Village to Revitalize a Community," *San Antonio Express-News*, November 28, 2010.

30. James Alfred Ball interview.

31. Conservative estimate is based on Ball family reunion committee members James Alfred Ball and Jozette Ruffin.

32. San Antonio is the county seat of Bexar County. The county is surrounded by Medina County, Bandera County, Kendall County, Comal County, Guadeloupe County, Wilson County, and Atascosa County.

33. *Social Explorer*, s.v. Tables, Census 1970, Tract 1814.

34. *Censusviewer.com*, s.v. Leon Valley; *Social Explorer*, s.v. U.S. Demographic Maps, Census 2000–2010.

35. For more on the megachurch, see Tamelyn N. Tucker-Worgs, *The Black Megachurch: Theology, Gender, and the Politics of Public Engagement* (Waco, TX: Baylor University Press, 2012), 21–50; and "The Black MegaChurch" (interview), *On Common Ground*, www.blogtalkradio.com/ocg/2011/10/30/our-common-ground-the-black-mega-church.

36. Rae Lowry interview.

37. According to Rae Lowry, another way that members were drawn to St. John's was its radio advertisement. Lowry interview; Antioch Missionary Baptist Church, "Antioch Missionary Baptist Church History," http://antiochsat.org/history.htm; Antioch Missionary Baptist Church, "Antioch Missionary Baptist Church History"; Mount Zion First Baptist Church, "History," www.mountzionfbc.org; New Light Baptist Church, "Church History," http://nlbcsa.org/churchhistorycont.html. *Social Explorer*, s.v. Tables, Census 2000, R10389514; Lacy, *Blue-Chip Black*, 21–50. Most black San Antonians who refer to themselves as the black middle class

in 2000 had a household median income of $35,000–$40,000. Based on sociologist Karen Lacey's conceptualization of the black middle class used in this article, most blacks surveyed thought that $50,000 made up what she called the core black middle class, and that people within the $35,000–$40,000 range were actually part of the lower middle class. My assessment of this group is that they are transitional people who see themselves as part of both the middle and working classes but have more in common with the working poor, making them part of what I call "the new working class."

38. *Social Explorer*, s.v. Tables, Census 1980–2010; *Censusviewer.com*, s.v. Converse, Kirby, Live Oak, Universal City, and Windcrest; see also *Censusviewer.com*, s.v. Cibolo, Garden Ridge, Schertz, and Selma.

39. Rae Lowry interview.

40. Rae Lowry and Herbert Ruffin I interviews. See also Anna Victoria Wilson and William E. Segall, *Oh, Do I Remember! Experiences of Teachers during the Desegregation of Austin's Schools, 1964–1971* (Albany: State of New York University Press, 2001).

41. For more on Southwest sensibilities, see Neil Foley, *The White Scourge: Mexicans, Blacks, and Poor Whites in Texas Cotton Culture* (Berkeley: University of California Press, 1999).

42. Rae Lowry interview.

43. Rae Lowry interview.

44. Lety Laurel, Rae Lowry, Meredith Elsik, Jessica Tamayo, James Shelby Ball, Jesse Ball, Honey Ruffin, and Ron Hall interviews. For information on the closing of Lackland AFB, see Ramiro Escamilla, "Kelly Air Force Base Contamination, a Dying Issue," *San Francisco Examiner*, July 24, 2009; and "Kelly Air Force Base: San Antonio's Dumping Ground," March 17, 2000, *Toxic Texas*, www.txpeer.org/toxictour /kelly.html.

45. *Social Explorer*, s.v. Tables, Census 1980–2010.

46. Jesse Ball interview.

47. Haya El Nasser, "Gated Communities More Popular, and Not Just for the Rich," *USA Today*, December 16, 2002, http://usatoday30.usatoday.com/news/nation/2002 -12-15-gated-usat_x.htm.

48. El Nasser, "Gated Communities."

49. "Mayors of San Antonio," *City of San Antonio*, www.sanantonio.gov/clerk /Archives/mayors1.aspx); Rae Lowry, James Ball, Jesse Ball, James Alfred Ball, Herb Ruffin, Lety Laurel, Meredith Elsik, and Jessica Tamayo interviews; Tedesco, "As S.A. Grows"; Melissa Ludwig, "Colleges Mirror S.A.'s Growth," *San Antonio Express-News*, January 10, 2008.

50. Lety Laurel interview. See also Dwonna Goldstone, *Integrating the 40 Acres: The Fifty-Year Struggle for Racial Equality at the University of Texas* (Athens: University of Georgia Press, 2012).

51. Tedesco, "As S.A. Grows"; Ludwig, "Colleges Mirror S.A.'s Growth."

52. Rae Lowry, James Shelby Ball, Jesse Ball, James Alfred Ball, and Doraine Ruffin interviews.

53. Jozette Ruffin, Rae Lowry, and James Shelby interviews. For more on the national gentrification movement of America's downtowns and where young professionals have been living, see Haya El Nasser, "American Cities to Millennials: Don't

Leave," *USA Today*, December 4, 2012, www.usatoday.com/story/news/nation /2012/12/03/american-cities-to-millennials-dont-leave-us/1744357.

54. Lisa Hsiao, "Project 100,000: The Great Society's Answer to Military Manpower Needs in Vietnam," *Viet Nam Generation* 1, no. 2 (1989): 14–37; Wallace Terry, *Bloods: Black Veterans of the Vietnam War: An Oral History* (New York: Ballantine Books, 1985); Robert McNamara, "Memorandum for the President: Subject: Project One Hundred Thousand," July 25, 1967, *African-American Involvement in the Vietnam War*, www.aavw.org/protest/draft_100000_abstract15.html; and Office of the Assistant Secretary of Defense: Manpower and Reserve Affairs. *Project One Hundred Thousand: Characteristics and Performance of "New Standards" Men* (Washington, DC: The Office, 1969) (OCLC Accession No. 09461968.)

55. Herb Ruffin interview; Herbert G. Ruffin II, "East Palo Alto (1925–)," *Blackpast. Org: Remembered and Reclaimed*, www.blackpast.org/?q=aaw/east-palo-alto-1925.

56. "San Antonio Market Trends," *Trulia*, www.trulia.com/real_estate /San_Antonio-Texas/market-trends), and similar charts on the website for North Castle Hills and East Palo Alto. In East Palo Alto the median cost for a four bedroom house in 2012 was $355,000; in 2008 it was $495,000. In San Antonio that same home in 2012 was $130,800, whereas in 2008 it was $97,650, a sign that the Alamo City's housing market is becoming more expensive. See also Performance Urban Planning report "8th Annual Demographia International Housing Affordability Survey: 2012," *Demographia*, www.demographia.com/dhi.pdf, 28, 31; according to this study, whereas the median price for a home in San Antonio is $156,200 and the median household income is $50,800, in the Bay Area the median price for a home ranges from $491,900 to $587,500 and the median household ranges from $73,800 to $84,900.

57. U.S. Bureau of the Census, *1970 Census of Housing: Volume 1, Housing Characteristics for States, Cavities, and Counties; Part 6, California* (Washington, DC: U.S. Government Printing Office, 1972), 7–10, 14–15, 453; U.S. Bureau of the Census, *1990 Census of Population and Housing, San Jose PMSA*, 1288–1289.

58. Meredith Elsik and Jessica Tamayo interviews.

59. In this closing section, my analysis borrows from sociologist Karyn Lacy's "Defining the Black Middle Classes Today," in *Blue-Chip Black*, in which the black middle class is defined by income, occupation, education, and wealth.

60. William A. Darity and Samuel L. Myers, *Persistent Disparity: Race and Economic Inequality in the United States since 1945* (Northampton, MA: Edward Elgar, 1998), 14–42; James Shelby Ball, James Alfred Ball, Jesse Ball, and Honey Ruffin interviews.

61. *Social Explorer*, s.v. Tables, Census 2000, R10389514. In San Antonio most blacks are part of the working class. In 2000 they made a median income of $29,598, which was comparable to $30,468 for Latinos/as and dissimilar to the $39,472 for whites.

62. Performance Urban Planning report, "8th Annual Demographia International Housing Affordability Survey," 28.

63. For a rich discussion of ring suburbs, or what many scholars call "black suburban boom towns," in areas such as Warrensville Heights (Ohio), Wellston (Missouri), Harvey (Illinois), East Palo Alto (California), Suitland (Maryland), and Roosevelt (New York), see Wiese, *Places of Their Own*, 215–17.

64. See Kenneth T. Jackson, *Crabgrass Frontier: The Suburbanization of the United States* (New York: Oxford University Press, 1985); Dolores Hayden, *Building Suburbia: Green Fields and Urban Growth, 1820–2000* (New York: Vintage Books, 2003); Becky Nicolaides and Andrew Wiese, eds., *The Suburb Reader* (New York: Routledge, 2006); Kevin M. Kruse and Thomas J. Sugrue, eds., *The New Suburban History* (Chicago: University Of Chicago Press, 2006); and David M. P. Freund, *Colored Property: State Policy and White Racial Politics in Suburban America* (Chicago: University of Chicago Press, 2007).

{ 10 }

FROM CHINATOWN TO LITTLE SAIGON

THE DEVELOPMENT OF A VIETNAMESE
ETHNIC URBAN CENTER IN HOUSTON

SON MAI

As Vietnamese refugees were settling along the Texas Gulf Coast after the conclusion of the Vietnam conflict, another phenomenon was taking place nearby in Houston, the largest city in the region. In late 1975 an interesting development began to appear east of downtown Houston, along Chartres Street. At the time, several new businesses began to open in this area of Houston that had previously been regarded as an economically depressed "white flight" area of Houston. These businesses were selling Asian groceries and delicacies such as *bánh chúng* and *mứt sen*; new restaurants sold entrees such as *hủ tiếu* and *phở*; barbershops and clothing stores began serving a new clientele—the Vietnamese, a group that had not existed in the area only a year before.[1] This Vietnamese shopping area, as well as the community around it, soon named "Vinatown," was the first of a new type of Asian ethnic community, deemed by sociologist Wei Li in 2008 as the "ethnoburb," or ethnic suburb.[2]

As Vietnamese refugees settled across the Gulf region, Houston became the urban center as well as the beacon for trade, entertainment, news, and the preservation of Vietnamese culture. As later groups of immigrants began to arrive to Texas from Southeast Asia, this large Vietnamese ethnic enclave often became the first destination before settlement in hinterland areas, such as the Gulf fishing communities. This Vietnamese ethnoburb community in Houston, often referred to as "Little Saigon," eventually became the model for most Asian community development across Texas, as well as in several areas across the country, differentiating itself from a traditional "Chinatown." These ethnoburban enclaves reflect the new post-1964 immigration, which included nearly all Vietnamese immigrants, as well as new attitudes toward East Asian immigration to the United States. Although there is a common Asian heritage, these ethnoburbs are unlike the older Chinese communities established along the Eastern Seaboard and West Coast during the late nineteenth century.

The first Asian ethnic communities in the United States—China-towns—were established in densely populated inner-city areas such as those in San Francisco and New York City. These ethnic enclaves were the entry points for the original immigrants from Guangdong Prov-ince, outside of Hong Kong, in the late 1840s as Chinese immigration to the United States began to take off, following social upheavals from the British victory in the opium wars, the decline of the Qing Dynasty, famine, banditry, and rumors of a "Gold Mountain" in California.[3] In San Francisco, the Chinese were initially welcomed for their work ethic and allowed to settle in any part of the city to open up businesses, leading many to choose a central part of the city to hawk their services and wares. In July 1863, in the midst of the American Civil War, mounting evidence began to appear that the Chinese were wearing out their welcome, which led to nativist attacks on New York's Chinatown during the race-driven draft riots of that year. Matthew Frye Jacobson maintains that this was because the Chinese were increasingly being seen as a "modification of the negro," as foreigners, and therefore not able to be accepted as full Americans.[4]

The Anglo American perception of East Asians did not improve after the Civil War. With the completion of the transcontinental railroad in 1869 during the Reconstruction era, thousands of Chinese—who had ful-filled their service contracts with the Central Pacific—found themselves unemployed and flooded the job market with more laborers than positions available. This trend culminated in a drop in wages that was compounded in the economic depression in 1873, leading to a rise of nativist sentiment, particularly against Asian immigrants. By this time, Chinese immigrants had banded together by living in the same boardinghouses as a form of protection. The anti-Chinese movement culminated in a race riot in San Francisco in 1877, which killed four people and caused $100,000 in dam-ages.[5] In the aftermath of this riot, the Chinese soon felt that strength in group numbers was a necessary way to defend themselves, their property, and their livelihoods against attacks by anti-Asian nativists. In the end, as in most parts of the world, the Chinese diaspora formed the original Chinatowns with the common purposes of using strength in numbers as a way to stave off persecution and violence as well as to ensure economic and cultural survival in an otherwise hostile environment.

About the same time that Chinese immigration in California was growing fast, a man named Ah Ken arrived in Manhattan to open up a smoke shop, rolling and selling cheap "nickel seegars" to working-class individuals while operating a boardinghouse in the upper-floor rooms of his business, used to rent out to other Chinese immigrants to augment his business income. Later Chinese immigrants, emulating the successes of Ah Ken, began to follow in his footsteps and opened up cigar shops as well but soon deviated by selling other sundries and general merchandise. As the numbers of unemployed railroad workers increased after the completion of the transcontinental railroad, many Chinese began to open similar shops in Manhattan, adding laundries, restaurants, and professional offices to form a business enclave that was also ethnically Chinese. As anti-Chinese hostilities and persecution arose after Reconstruction, the Chinatown district grew as more migrants settled in the area as a form of protection by numbers. By 1880 there were around two thousand Chinese residents along an approximately ten-square block area.[6]

Away from the bourgeoning metropolises of New York and San Francisco, the first Chinese contract workers arrived in Texas in 1870 to work on the railroads. Several men who completed (or broke) their contracts with the Southern Pacific and the Houston and Texas Central Railway companies left the state to find better opportunities in the big cities along the Atlantic and Pacific coasts—but many stayed and sought work in agricultural occupations or business ventures. The Chinese in Texas initially settled in railroad centers such as Fort Worth, El Paso, and San Antonio, opening businesses such as general merchandise stores, barbershops, boardinghouses, restaurants, laundries, and opium dens.[7] In Houston there was a fledgling Chinese community on the east side of downtown that coexisted in a predominantly black neighborhood and sold merchandise and services to the poorer part of town, which most Anglo merchants shunned because of the smaller profit margins.[8] Meanwhile, in Fort Worth the Chinese settled in the city's slum district known as "Hell's Half Acre" and formed what the *Fort Worth Daily Democrat* deemed in 1878 as a "Little Chinatown," with about sixty residents by the turn of the century, complete with boardinghouses, restaurants, hand laundries, and opium dens—the latter also catering to an Anglo American clientele and providing the city's residents the only means to obtain the illicit drug.[9] Despite activities in these railroad towns before the turn of the century, none of the early Asian settlers in Texas formed a true ethnic community but rather worked as middlemen and mediators between blacks and whites in a post-Reconstruction Jim Crow society. There was no need for a true ethnic enclave at the time, for Asian population numbers in the state remained low, and the Chinese did not constitute enough of an economic threat to necessitate isolation for protection.

Ultimately, nativist sentiment against the Chinese was significant enough to gain Congress's attention, leading to the passage of the Chinese Exclusion Act in 1882. This severely curtailed Asian immigration and had many social consequences in the nation, including Texas. The fledgling Chinese population—as well as other Asian groups—were unable to construct ethnic enclaves, and their populations remained small until the end of World War II.

AFTER WORLD WAR II

A friendlier postwar environment in Texas made it possible for Asians to settle in the state and form new ethnic communities. During World War II, in recognition of the Republic of China's support for the Allied war effort, Congress formally lifted the ban on Chinese immigration in 1942 and allowed for a token one hundred Chinese immigrants to the United States each year under the National Origins Quota Act. However, it would not be until the Immigration Act of 1965, followed by the rise of the Texas "Sun Belt" economy, and the end of the Vietnam conflict that the state experienced large East Asian immigration that allowed an opportunity for the introduction of ethnic enclaves. These new Asian communities would, however, develop differently from the original inner-city Chinatowns, reflecting the postwar automobile culture as well as a different racial climate. Sociologists have termed these new communities "ethnoburbs"—since they tend to be ethnic communities outside urban centers and share many characteristics with suburban America.

During the height of the civil rights movement in the early 1960s, which brought gains for African Americans, particularly by reducing segregation and unfair voting practices, President Lyndon B. Johnson's administration also sought ways to end discrimination in other groups, including immigrants, as part of a program dubbed "The Great Society." As the Johnson administration began to scrutinize previous practices for unfairness, the

National Origins Quota Act, established in the 1920s, became regarded as an obsolete and unfair way of allowing entrance to the United States, with preference given to those of northern and western European backgrounds and restrictions on immigration from Africa and Asia. The Immigration and Nationality Act of 1965 would abolish the quota system and replace it with a new formula that placed preference on granting immigrant visas to those with marketable skills and with spouses or family members who were U.S. citizens who could act as sponsors upon arrival. The first criterion proved to be important in the development of sustainable Asian communities in Texas, as the state's postwar economy quickly moved toward high-tech manufacturing at the same time that several East Asian countries—notably South Korea, Hong Kong, Singapore, and the Republic of China (Taiwan), the "Four Asian Tigers"—also experienced rapidly developing economies, moving from agricultural to heavy industry and high-tech production and utilizing an educated labor force that was also desirable in the United States.

In light of the cold war and the Soviet launch of Sputnik, which exposed American weakness in science, technology, engineering, and mathematics (STEM) fields, the United States encouraged Asian immigration from these "Tiger" economies, as well as Japan, to take advantage of the 1965 immigration legislation and help boost the number of educated scientists, engineers, and technicians in cold war industry. This came with the hope that the United States, already beginning to lag in STEM fields, would receive an influx of skilled immigrants to keep up with the cold war communist bloc. According to Ivan Light and Edna Bonacich, "The importation of skilled foreigners was the counterpart to domestic programs intended to support education and training of the native born."[10]

After Jack Kilby first designed the integrated circuit in 1958 while working for Dallas-based Texas Instruments (TI), the state became a leader in producing solid-state electronics, leading to the eventual phase-out of vacuum tube technology. Ultimately, labor costs in the East Asian "Tiger Cub" economies of Indonesia, Malaysia, Thailand, and the Philippines would drop to the point where semiconductor manufacturing operations would move to the Far East, taking advantage of low costs and an educated workforce there. But for the next forty years, following TI's lead in microchip research, development, and manufacturing in Texas, other technology firms such as Fujitsu, Raytheon, and National Semiconductor began operations in the state. A series of spinoffs from TI formed other

Texas-based firms such as Mostek and Cyrix, which were bought out by European-based Thompson and Taiwan-based VIA Technologies, respectively, by the end of the century. At the same time, local Texas-based tech firms also opened up in other parts of the state, notably Compaq Computers in Houston followed by Dell in Austin.

The state's rapidly moving technology and information services industries were synchronous with the rise of technology giants Sony, LG, and Samsung across the Pacific in East Asia. Improvements in communications and transit networks connected Texas with other regions of the globe, and international trade grew significantly, especially after the opening of two major world-class air hubs at Dallas–Fort Worth International and Houston-Bush Intercontinental by the late 1970s. This placed Texas in a good position for globalization through foreign trade, and with it cultural and population exchanges. By 1980 the U.S. census reported that approximately 80,000 non-Indochinese East Asians had settled in the state since 1970.

REFUGEES AND THE SUBURBS

While the Asia Tiger and Tiger Cub economies were growing along with the Texas economy, other parts of the world struggled with turmoil. By 1975 political and military conflicts in Southeast Asia led to the rise of victorious communist regimes in Cambodia, Laos, and Vietnam. In April of that year, Americans watched in shock as U.S. Marines evacuated fellow citizens and vulnerable Khmer in Operation Eagle Pull, as the genocidal Khmer Rouge, led by Pol Pot, advanced on Phnom Penh. About a month later, Americans once again watched frantic U.S. Marines hastily evacuate over 7,000 citizens and nearly 135,000 desperate South Vietnamese to safety as the besieged capital city, Saigon, was being overrun by the communist North Vietnamese. Since most Indochinese evacuees did not qualify for normal immigration channels, the Ford administration passed the Indochina Migration and Refugee Assistance Act, which empowered the U.S. attorney general to give refugee status and parole to the Indochinese. By 1980 the census reported that approximately 40,000 Vietnamese were resettled in Texas; their numbers were too insignificant to be recorded in the census only a decade earlier.

The large Asian populations in the state necessitated the creation of ethnic communities by the 1980s, to cater to a diverse population that ranged from educated professionals to unskilled refugees. The new Asian communities in Texas developed into a new type—one that was formed not for protection against nativist and racial attacks but rather for economic activity with an emphasis on cultural preservation adapted to Texas's new postwar automobile culture. The result was a formation of a new type of ethnic community—one not located in a densely populated inner-city area but rather following on the suburban development and hinterland sprawl common in the post-1960s Sun Belt. Not only were these communities formed to cater to the car culture, but they also sought to dispel many of the stereotypes that arose from the old Chinatowns of the Northeast and West. Rather than spontaneous growth in inner-city areas with narrow, dark alleys, perceptions of vermin, underworld crime, and illicit activities, the new Asian communities were away from downtown areas, with wide boulevards, strip malls, ample parking lots, community police liaisons, and in some cases even former "big box" retailers repurposed as enclosed malls, cultural centers, and houses of worship. These new ethnoburbs formed in suburban areas in large cities throughout the South, including the Dallas–Fort Worth area, Houston, and Austin.

The first ethnoburbs in Texas formed to serve the growing Chinese and Taiwanese professional communities, such as in Houston, home to computer giants Compaq and CISCO as well as technologically intensive industries such as defense, dominated by Haliburton, and a plethora of growing energy firms. The Asian ethnoburb movement in Houston occurred at the right place and the right time; a real estate bubble—compounded by political scandal—brought a windfall for the city's Asian American population and led to the development of the state's largest Asian ethnic enclave. This real estate boom and bust had begun in 1955, when Houston land baron Frank Sharp purchased a large tract of land in the southwestern portion of town, egotistically named Sharpstown, which he envisioned as becoming a master-planned community, complete with Jim Crow–era racially based deed restrictions that would keep most minority groups out. In 1961, Sharp constructed the centerpiece of his new suburban utopia, the Sharpstown Mall (now PlazAmericas), Houston's first fully enclosed and air-conditioned shopping center. In the center of the mall was the 114,000-square-foot, ten-story Sharpstown State Bank building (now the Jewelry Exchange Center), with the penthouse floor allocated for Sharp's office so he could overlook his fledgling fiefdom, which contained several strip malls, apartment complexes, suburban style housing, and a Jesuit college preparatory school.

Sharp's empire crumbled in 1971 in the wake of the Sharpstown scandal, a "pump and dump" scheme that involved his providing insider information to at least thirty Texas state legislators and encouraging them to purchase stock in Sharp's insurance business, creating artificial demand and, accordingly, artificially inflated stock values.[11] Though Sharp himself was sentenced to only three years' probation and a fine, the Sharpstown scandal effectively ruined the political careers of several state politicians, including Governor Preston Smith. At the same time, the scandal opened the way for the Republican Party, suffering since the Reconstruction era, to make a comeback in the state. More relevant here, the scandal resulted in investors pulling out of the Sharpstown project, and by the early 1980s the area faced declining property values, run-down apartments, and the effects of white flight as the affluent gave up on southwest Houston for the new suburban developments in the northern and western parts of the metropolitan area—toward The Woodlands and Katy, respectively. These declining property values, compounded by a land bust in the mid-1980s, made it possible for Chinese American investors to purchase parcels of distressed real estate and redevelop them into an ethnoburb to service the town's growing Asian American community. The Sharpstown's Asian community grew even further when another concept—gentrification—began taking place with the redevelopment of an older Chinese and Vietnamese community just outside downtown Houston into luxury condos.

As the Asian populations—mainly Chinese and Taiwanese—grew in these early ethnoburbs, business and commerce naturally followed to cater to the bourgeoning ethnic communities. These businesspeople and developers entered at the right moment. With real estate prices depressed as affluent Anglo Americans moved to newer developments, Asian real estate investors bought struggling strip malls, abandoned "big box" retail outlets, and convenience stores, redeveloped them, and marketed these retail and professional spaces to other immigrants for lease as small shops and offices. Adding to businesses, several religious missions opened worship centers to serve the immigrant communities. With a growing population eager to purchase and receive food, products, and services from the old country, a business community followed to cater to this new Asian population, complemented by cultural activities to form a complete ethnic community.

As U.S.-based refugee camps closed in December 1975, small Vietnamese grocery stores came into existence in Houston on the southern outskirts of downtown, in an area known as Midtown. The Midtown Vietnamese enclave would span an eighteen-block area dominated by the Hoa Bình (Peace) Shopping Center and eventually eclipsed the initial Vinatown to take over as the central area for Vietnamese life in the Gulf Coast region. The Hoa Bình Shopping Center was soon supplemented by ethnic restaurants, barbershops, law firms, insurance agencies, and other service-based businesses. Within a decade, this clustering of shops and professional offices evolved into a formidable "Little Saigon" ethnic enclave and became the center of Vietnamese life in the region. By 1998 the Little Saigon in Midtown had become a bustling community, complete with supplementary Vietnamese street signs affixed under English signs. These streets were subtitled with the names of major thoroughfares from pre-1975 Saigon: Tran Hung Dao Street, Le Loi Street, Hai Ba Trung, and Tu Do Boulevard.

By 2003, however, the Midtown Little Saigon was undergoing rapid deterioration and decline, fostered by highway construction projects in the area as well as urban redevelopment sparked by gentrification. Ultimately, rising property values would outprice most Vietnamese residents and businesses in the area. By 2005 the Hoa Bình Shopping Center was undergoing demolition to make way for a Spec's Fine Wine shop. Most of the buildings in the area have since been razed to construct luxury downtown lofts for young professionals and their families. There are a few traces remaining of the Vietnamese community in the form of restaurants, which now serve a more pretentious foodie and hipster clientele rather than a working-class immigrant community.

The downtown Vietnamese communities failed because the original purposes for an inner city and dense ethnic enclave were no longer applicable in the post-1965 racial landscape. Though race continued to play a large role in the development of these ethnic communities, the new Texas ethnoburbs were created with the opposite effect. Now, a "model minority" stereotype prevailed and, as white flight became a reality in many communities, city leaders were inclined to welcome Asian American settlement and business development, with the impression that these groups were more affluent and less likely to engage in criminal activities that had plagued white flight neighborhoods dominated by other minority groups, mainly African Americans and Hispanics.[12] These leaders perceived Asians as a population that could be rapidly assimilated into the general population and become citizens of quiet demeanor.

As early as 1975, this rapid assimilation theory gained currency when the federal government introduced a "dispersion policy" to resettle 125,000

Vietnamese refugees across the country, with the intention of preventing an ethnic enclave not unlike the Cuban American–dominated Little Havana in Miami from forming. Not surprisingly, these civic leaders who subscribed to the "model minority" image of Asians did not believe that social issues pertaining to every group also applied to Asian immigrants. Nevertheless, by the early 1990s many of these communities struggled with problems that are endemic in virtually every town across the United States: immigrant adults working long hours to make ends meet, parents unable to supervise their children adequately, which leads to the latter struggling in school and becoming involved in everything from petty theft to organized crime.

Soon, Asian American developers were taking advantage of the depressed property values and revitalized depressed neighborhoods in other parts of the state, as well as the nation, to develop similar ethnoburbs. For example, in Arlington, Texas, the central part of town was facing the threat of urban decay, as old strip malls like the French Quarter on Pioneer Parkway that once catered to working- and fledgling middle-class Anglo American families with Winn Dixie, Woolco, Revco, Mott's Five-and-Dime, and Baskin Robbins began to show their age, decline, and signs of decay. This area became the Phước Lộc Thọ Asian Market Center, anchored by a Hong Kong Supermarket, an Arc-En-Ciel (Rainbow) Vietnamese restaurant, at least two eateries selling *phở*, several bakeries, barbershops, and even a money remittance center to send cash to families back in Asia. By the 2000s, Asian American redevelopers had converted at least two other declining strip malls along Arlington's Pioneer Parkway corridor to become booming shopping areas in this new ethnoburb. Furthermore, during the mid-2000s construction boom, developers constructed at least two more shopping centers from the ground up, and though the bust and following recession slowed down new construction, the pace of converting and upgrading preexisting strip malls continued during the economic downturn, since the Asian developers' business model depended on poor economic conditions to "buy low" as a long-term investment strategy.

VIETNAMESE CULTURAL PRESERVATION

In any thriving ethnic community, jobs and investment potential are not the only factors keeping the neighborhood alive; a shared cultural element is required to sustain new ethnoburbs. This begins with the introduction of a community center, likely a church or a temple, followed by other forms of communication such as a newspaper. In the more dispersed car-based ethnoburb in Texas, where not everyone has access to reliable transportation, new media becomes the most popular and effective way to spread information and keep the community informed about news and events within the community, as well as in the larger metropolitan area and in other ethnoburbs across the region.

The religious community was usually the first point of cultural activity and information within the early ethnoburbs. Faith-based institutions also played a significant role in the preservation of Vietnamese culture in Texas. Initially, white-dominated churches sponsored Vietnamese congregations and allowed the communities to share facilities and offer services and Sunday school classes in their native tongue. The concept of bilingual churches in Texas did not, however, appeal to the extremely independent-minded Vietnamese communities, which desired to preserve their culture through voluntary separation. Whereas other ethnic groups, such as Hispanics and Koreans, largely preserved their links with their English-speaking counterparts by continuing the sponsorship concept, the Vietnamese living in Texas believed that independent congregations were necessary to maintain their own traditions. This belief resulted in an active movement to form separate houses of worship. Within two years of initial arrival in the United States, Vietnamese Buddhist temples began to appear. In that same year, 1977, the first Vietnamese Christian congregation in Texas, Queen of Vietnam Catholic Church, opened in Port Arthur. Soon other Vietnamese communities, in Houston as well as the fishing regions, followed suit by constructing separate church buildings.

Vietnamese soon began constructing congregations, parishes, and temples in Asian architectural styles to serve religious adherents. In Houston, as well as in other ethnoburban communities, however, many of these church structures were in former big-box retailer buildings that served an Anglo American suburban community before they closed with changing demographics. The Vietnamese Martyr's Catholic Church in Arlington, for example, was once a Food Lion supermarket that sat abandoned for several years until it was redesigned by Vietnamese American architect Khiet Nguyen and repurposed into a two-thousand-seat parish by 2000. Seven years later, the facility underwent further remodeling. Today, no trace of a big-box retailer remains; the sprawling church campus has been extensively refashioned to reflect the ethnoburban spirit of renovation.

As newspapers began to give way to radio, the internet, and television as main sources of information, the Asian American ethnoburbs took advantage of these new media resources to spread news and other information. For example, in the Vietnamese community in Houston, two AM radio stations had come on the air by 2000. With the transition from analog television to digital in the mid-2000s, the Federal Communications Commission effectively created new opportunities to set up subchannels within a frequency band, allowing ethnic broadcasters to offer more foreign-language programming without the substantial capital that traditional ethnic stations like Univision had to raise in the 1980s. By 2010 there were two television stations serving the 135,000-strong Vietnamese American community in Houston, and one station for 110,000 people in the Dallas–Fort Worth area, broadcasting community news as well as cultural programming twenty-four hours a day.

Since the late 2000s these Asian ethnoburbs in Houston and Dallas–Fort Worth have begun to face a new challenge: the earlier developers and residents of these ethnic enclaves are moving upwards and, with rising affluence, have begun moving to other suburban neighborhoods. In Houston, for instance, some of the earliest residents of the Asian community in Sharpstown have emulated their white predecessors and moved southwest, heading toward Sugar Land, attracted by better schools for their children, improved housing standards, affordable costs, and easy access to the old neighborhoods by car. Once again, city leaders welcome this type of economic development, since it brings in a tax base and, above all, minimizes urban decay and blight.

CONCLUSION

As observed with recent phenomena in the Houston and Dallas–Fort Worth areas, the Asian ethnic enclaves in Texas developed with different reasons than the traditional Chinatowns in U.S coastal cities. The Texas ethnoburbs developed not as a form of protection against a racially hostile host country but more as a way to serve an immigrant community by providing goods, services, news, information, and culture from home by taking advantage of a suburban neighborhood's changing demographics. As Bonnie Tsui has described the creation of a Chinese ethnoburb in Las Vegas after a major real estate bust there in *American Chinatown*, "If you build it, they will come . . . by car."[13] The reasons for the rise of the ethnoburbs are clear, but a question about the future remains: Will these ethnoburbs be sustainable, or will increased property values from redevelopment result in owners cashing out and redeveloping other areas? The original Houston Chinatown, once located near downtown before moving to Sharpstown, is an example of the fluidity of the new Asian ethnoburb in Texas, and only time will tell if these communities have a permanent presence in the community landscape.

NOTES

1. John K. LeBa, *The Vietnamese Entrepreneurs in the U.S.A.: The First Decade* (Houston: Zieleks, 1985), 113.

2. Wei Li, *Ethnoburb: The New Ethnic Community in Urban America* (Honolulu: University of Hawaii Press, 2008), 2.

3. Bonnie Tsui, *American Chinatown: A People's History of Five Neighborhoods* (New York: Free Press, 2009), 7–8.

4. Matthew Frye Jacobson, *Whiteness of a Different Color* (Cambridge, MA: Harvard University Press, 1998), 53–54.

5. Iris Chang, *The Chinese in America: A Narrative History* (New York: Viking Press, 2003), 127–28.

6. Min Zhou, *Chinatown: The Socioeconomic Potential of an Urban Enclave* (Philadelphia: Temple University Press, 1992), 33–35.

7. Marilyn Dell Brady, *The Asian Texans* (College Station: Texas A&M University Press, 2004), 15.

8. Fred von der Mehden, ed., *The Ethnic Groups of Houston* (Houston: Rice University Press, 1984), 67–68.

9. Richard Selcer, *Hell's Half Acre* (Fort Worth: Texas Christian University Press, 1991), 222–24.

10. Ivan Light and Edna Bonacich, *Immigrant Entrepreneurs: Koreans in Los Angeles, 1965–1982* (Berkeley: University of California Press, 1988), 133.

11. Robert A Calvert, Arnoldo de Leon, and Gregg Cantrell, *The History of Texas*, 3rd ed. (Wheeling, IL: Harlan Davidson, 2002), 419–20.

12. Lori Rodriguez, "Census Tracks Rapid Growth of Suburbia," *Houston Chronicle*, March 10, 1991, sec. A, 1.

13. Tsui, *American Chinatown*, 204–5.

CONCLUSION
AN INVITATION FOR FURTHER RESEARCH

M. SCOTT SOSEBEE

When Hurricane Harvey dumped its massive amount of rain on the Texas Gulf Coast, the results devastated the Houston metropolitan area. The highest totals in the Houston metropolitan region neared fifty inches, and meteorologists estimated that the sheer volume of water that fell on the region was in excess of 25 trillion tons. The Houston Independent School District, which is the nation's seventh largest, indefinitely suspended all classes, which affected more than 200,000 students. Some estimates predicted that at least 30,000 people would be, at least, temporarily homeless. The scale of the disaster was mind-boggling.[1]

What, you may ask, does a hurricane in the urban center of Houston have to do with Texas's suburbs? After reading the preceding chapters the answer should be clear. The pattern of growth within the Houston metro area in the past five decades, as in Texas's other urban areas, has generally followed the pattern of the suburban process. Growth in Houston, especially since 1970, did not occur in the central urban core but in the decentralized margins. The economic activity—shopping centers, housing developments, and multilane freeways and other transportation conduits—moved to the outer ring of the city. People followed those new expansions into the suburban areas around Houston.[2] The statistics of the growth are amazing. Harris County's population in 1970 was 1,709,436, and in 2016 it was estimated to be 4,598,928. That is a net gain of 2,880,492, an increase of 168 percent. From 2010 to 2016 alone the Houston–The Woodlands–Sugarland metropolitan statistical area (MSA) added 851,971 new residents, the greatest increase for any metropolitan region in the nation. Put another way, the MSA absorbed the population of Baton Rouge, Louisiana, in just six years. Yet population statistics tell only half the story. Harris County is home to 1,768,827 housing units, second only to Los Angeles County in the United States. It lists more than one hundred thousand registered "employer establishments" and economic output of over $600,000,000.00.[3]

What such growth means is that tons and tons of concrete, steel, rock, and other materials transformed what were once open green spaces and rural homesites into what during Hurricane Harvey became massive ribbons of raging rivers churning with water, roiling through homes, businesses, and roads. But Houston's flooding was not the result of a hurricane's storm surge or the breaching of a levee. Rather, it was to a large extent an event that was man-made, one that was a direct result of the massive suburban sprawl that is characteristic of almost all of the Texas urban areas.[4]

Houston lies in an area that receives an abundance of rain, and its geography makes it prone to temporary flood conditions. However, for decades—even more than a century—the rain that fell was absorbed by the region's equally abundant grasslands and other foliage. Certainly, large amounts of rain over a short period can cause a flood even when nature utilizes every mechanism it has to block such an event. And Houston has had its share of flood events since its founding, events that certainly caused devastation within the city but were attributable almost wholly to natural phenomena and not necessarily human intervention. The city constructed flood protection beginning in the 1930s and withstood a number of huge rain events in the next fifty to sixty years, but the effectiveness of those flood prevention constructions could not help Houston avoid massive flooding in the late twentieth and twenty-first centuries.[5]

The rapid suburbanization process eroded Houston's natural and human barriers to flood events. Ground absorption is the key to avoid cataclysmic floods; sometimes rain is too abundant for the ground to absorb, which creates runoff, which was the reason the Harris County Flood Control District built the Addicks and Barker dams. However, the most precise reason that runoff becomes a problem in the twenty-first century in large cities is that urban sprawl covers the ground with concrete and other structures. And Houston and its environs have a lot of concrete and other structures. Essentially, suburbanization created thousands of square miles of the exact construction that makes runoff multiply. The result? Massive floods, such as Houston experienced in 2001 after Tropical Storm Allison, 2008 after Hurricane Ike, and during numerous other rain events such as the storms that spawned flooding in 2003, 2006, and 2016. Then came Hurricane

Harvey in 2017. Perhaps, city, state, and federal leaders will explore actions that prevent other such disasters, although the vast suburban sprawl in Harris County may prevent any real preventative action.[6]

The results and problems of a hurricane-spawned flood may seem an odd opening to a conclusion on the suburban process in Texas, but in many ways it is apropos. Paul J. P. Sandul opens this volume with the assertion that Texas is, in fact, a suburban state, and the following essays make precise points about how suburbanization has shaped the nature of Texas since at least the mid-twentieth century, and also how it will continue to be one of the most intriguing forces in "molding" Texas and Texans as the twenty-first century progresses. Many Texans may have cultivated a narrative of living in a state that possesses a rural soul, with values and characteristics more applicable to a nineteenth-century frontier, but that is a constructed history designed to entrench an identity more than to reflect reality. The truth is that Texas—and Texans—more often than not do not resemble the stereotypical image of the state. Texas is an urban state; over 21,000,000 of its 28,000,000 people live in urban areas, which is second to California in the number of urban residents. In other words, 75 percent of its residents are urban dwellers.

Census data for Texas indicate that almost half the residents of the state's three largest cities (Houston, San Antonio, and Dallas) reside in suburban regions. Beyond such data, however, is the fact that when you analyze other large cities in Texas, places such as Austin, Corpus Christi, Lubbock, and Tyler, the pattern of development and form you find is suburban in nature. Lubbock is a city with no central core, with outward sprawl of multiple neighborhoods, shopping districts, and a pattern of growth that encourages even further spread. Thus, although Lubbock is an isolated city on the south Plains, it is suburban in nature, with culture, politics, and society that reflect just such a development. Tyler, 450 miles to the east of Lubbock, and occupying land in which conditions are almost polar opposite from its fellow Texas city on the south Plains, has an almost identical physical makeup. Tyler's downtown, like Lubbock's, is in no way the economic or social hub of the city. Rather, again like Lubbock, Tyler's vibrancy and development have developed on the fringes of the city, an extension that is almost a mirror of the prototypical suburban process. Yes, Texans are a suburban people.[7]

The twenty-first century has even exacerbated the growing tint of Texas as a suburban state. Texas's suburban population, which has been expanding for at least the past five decades, has grown even faster since 2000, with perhaps the most exponential growth occurring in the past decade. A pertinent example is Hays County, in the booming I-35 corridor between Austin and San Antonio. Hays County grew from 19,934 people in 1960 to 97,589 in 2000, a remarkable rise of 66 percent in forty years. Such a remarkable rise next gave way to one even more astonishing in the next decade and a half. The 2010 U.S. census counted 158,275 people in Hays County, a ten-year growth in population that was almost as much as the population gain over the previous extraordinary forty years of population increase. The trend does not appear to be lessening; current estimates peg Hays County's 2017 population at 201,739, which would be an increase of more than 25 percent in just seven years. It is now the fifth-fastest-growing county in the nation.[8]

Buda, Wimberley, and Dripping Springs—the primary towns in Hays County—were not too long ago small hamlets, but today they are suburban cities serving the booming metropolis that is Austin and San Antonio. Along with the population growth, the county has also seen growth in all that often comes with rising population and synthetic association with a large urban area. Property values have risen exponentially in the area, so much so that the former small farms that dotted the county have been replaced with upscale suburban housing subdivisions. As more affluent residents have moved into the environs of the county, working-class and traditionally middle-class neighborhoods have disappeared and consolidated, creating class tension that generally did not exist in Hays County decades before. Crime has also risen, and conflicts within the county school systems, discord that seemed to be present only in "big-city" schools just a short time before, have now found a home in Hays County.

Hays County is but an indicator of what may be in store for the future. Texas will encounter the burgeoning problems that confront suburbs all over the nation. Texas's suburbs—including those "island cities" whose development resembles the suburban process—are transforming and are no longer the exclusive havens of white flight and the well-to-do; certainly some still are, but as Theodore and Gwen Lawe detailed in their contribution to this volume, white flight is moving farther afield. Problems that have traditionally confronted only urban cores will begin to find a home in Texas's suburban reaches, issues such as racial tension, poverty, and political division.

The point is that, although the suburban process has been one of the most significant components of postwar Texas development, as the preceding chapters have proven, twentieth-century Texas's predominant pattern has been suburban in nature, which means that twenty-first-century Texas will have to confront the problems and transformations that come with being a predominantly suburban place. It is also a continuing process, one that will endure but also create new realities and different scenarios for the nation's second-most-populous state. The old social, cultural, and political patterns will change, and Texas's leaders and citizens will have to be ready to confront these changes as they appear. The articles in this volume go a long way toward introducing scholars and interested parties to the coming nature of Texas's urban areas and the citizens who live within them. We, the editors, conceived this work as a "conversation starter," a book that can serve as an introductory place for other studies and more in-depth looks at all parts of the suburban process in Texas. The authors of the volume have complied with the charge. We hope that it has accomplished its goal.

NOTES

1. The final official statistics from the National Weather Service and FEMA, in many cases, exceeded the estimates. The previous record for rainfall from a single cyclonic event on the continental United States was 48 inches (Tropical Storm Amelia in 1978), but precipitation from Harvey exceeded that total in a remarkable eighteen different spots. The greatest amount was 60.56 inches in Nederland in Jefferson County, and in the Houston metro area the most prolific location for rain was 51.5 inches in Mount Belvieu, forty miles east of downtown Houston. The human and material cost was even more disastrous. Through January 2018, 738,000 people had registered for aid from FEMA, 203,000 homes received damage and 12,700 were wholly destroyed, and sixty-eight people lost their lives as a direct result of the storm. United States, National Weather Service, Silver Springs, MD, "Hurricane Harvey Impact Report," January 23, 2018, 6–8, www.fema.gov/disaster/4332 (accessed January 29, 2018).

2. Because of Houston's unique annexation policies and growth patterns, not all of the "outer ring" areas of Houston are the traditionally defined separate municipalities of suburbs. For example, Kingwood is most definitely a suburb, a place born of the suburban process, but it is within the city limits of Houston.

3. For all census data cited in this Conclusion, see the tables in chapter 1. "Quick-Facts, Harris County, Texas, *United States Census Bureau*, www.census.gov/quickfacts /fact/table/harriscountytexas/BZA210216#viewtop (accessed on January 30, 2018).

4. "Houston's Flood Is a Design Problem," *Atlantic*, August 28, 2017, 45–57.

5. The city was flooded in 1838 when massive rain storms caused Buffalo Bayou and White Oak Bayou to leave their banks; other floods hit the city in 1929 and 1935. The 1935 storm was particularly devastating to the downtown area, which prompted the creation of the Harris County Flood Control District, which in turn contracted with the U.S. Army Corps of Engineers to build Addicks and Barker dams and reservoirs, which were specifically constructed to protect downtown Houston. And they worked; despite massive rainfall associated with hurricanes Audrey (1957), Carla (1961), and Alicia (1983), Houston suffered no catastrophic calamities.

6. United States Weather Service, Houston, Texas Historic Climatological Data and Events; "Houston's Flood Is a Design Problem," 47–49.

7. Becky M. Nicolaides and Andrew Wiese, eds., *The Suburb Reader* (New York: Routledge, 2006), 345–48.

8. "Suburban Population Continues to Surge in Texas," *Texas Tribune*, March 24, 2016, www.texastribune.org/2016/03/24/suburban-population-counties-surge-texas (accessed February 19, 2018); "County Population History," in *Texas Almanac*, 2004–2005 (Dallas: Dallas Morning News, 2004), 391.

CONTRIBUTORS

ANDREW C. BAKER is Assistant Professor of History at Texas A&M University–Commerce, where he teaches twentieth-century U.S., agricultural, and oral history courses. He is the author of *Bulldozer Revolutions: A Rural History of the Metropolitan South* (University of Georgia Press, 2018). His other publications include "Metropolitan Growth along the Nation's River: Power, Waste, and Environmental Politics in a Northern Virginia County, 1943–1971," *Journal of Urban History* 42 (January 2016); and "From Rural South to Metropolitan Sunbelt: Creating a Cowboy Identity in the Shadow of Houston," *Southwestern Historical Quarterly* 118 (July 2014).

ANDREW BUSCH is Assistant Professor of Interdisciplinary Studies in the Honors College at Coastal Carolina University. He is the author of *City in a Garden: Environmental Transformations and Racial Justice in Twentieth-Century Austin, Texas* (University of North Carolina Press, 2017). He is currently working on a project titled "The Myth of the Market: Texas and the Paradox of Neoliberal Economics," a study of the relationships among government, academia, and business in Texas since World War II.

ROBERT B. FAIRBANKS is Professor of History at the University of Texas at Arlington and a Fellow of the Center for Greater Southwestern Studies at UT Arlington. He has written various books and articles on the urban Southwest including *For the City as a Whole: Planning, Politics and the Public Interest in Dallas, Texas, 1900–1995* (Ohio State University Press, 1998) and most recently *The War on Slums in the Southwest: Public Housing and Slum Clearance in Texas, Arizona, and New Mexico, 1935–1965* (Temple University Press, 2014).

GWENDOLYN McMILLAN LAWE was a teacher in the Dallas Independent School District until 2011. She is the author of two books, *From Wolf to Wolfwood*, and *Taking Care of Mother*. She is a cofounder of the A. C. McMillan African American Museum in Emory, Texas (with her late husband Theodore), and is currently President of the East Texas Historical Association.

THEODORE M. LAWE was executive assistant to Detroit mayor Roman Gibbs and the first African American assistant to city manager George Schrader in Dallas. He left public service in the early 1980s to begin a career as an entrepreneur in Dallas. He also served as the curator of the A. C. McMillan African American Museum in Emory, Texas. He published articles in the *East Texas Historical Journal* and served as the first African American president of the East Texas Historical Association.

SON MAI is a native-born Texan interested in the study of Vietnamese diasporic communities. He received his B.A. in history from the University of Texas at Arlington, where he was a student of Robert B. Fairbanks; an M.A. from Stephen F. Austin State University; an M.B.A. from McNeese State University; and a Ph.D. from Texas Tech University. He is currently Instructor of History at McNeese State University in Lake Charles, Louisiana.

JAKE Mc ADAMS graduated with an M.A. in public history from Stephen F. Austin State University in 2013. During his time at SFA, McAdams received numerous awards including a Ben. H. Proctor Student Research Award and an Ottis Lock Research Grant. McAdams's research has focused on historic preservation and oral history. He currently lives in Granbury, Texas, and works as a consultant to small Texas cities on planning and infrastructure projects.

TOM McKINNEY holds a Ph.D. from the University of Houston and is currently the Director of Learning Resources at Angelina College in Lufkin, Texas.

PHILIP G. POPE is a banker in Lubbock, Texas, and teaches Texas history at Texas Tech University. He has published several articles and maintains research interests in Texas, the American West, community formation and identity, and small-town history. Pope received his Ph.D. from Texas Tech University in 2013.

HERBERT RUFFIN II is Associate Professor of History and Chair of African American Studies at Syracuse University. Ruffin holds a Ph.D. in American History from Claremont Graduate University, California. His research examines African American experiences in the twentieth- and twenty-first-century U.S. West, and in particular, social movements and the process of urbanization and suburbanization in the San Francisco Bay area and in Central Texas. Ruffin has published two books on the African American West with the University of Oklahoma Press: *Uninvited Neighbors: African Americans in Silicon Valley, 1769–1990* (2014) and *Freedom's Racial Frontier: African Americans in the West from Great Migration to Twenty First Century*, coedited with Dwayne Mack (2018). He has also published a book on African independence, *Illuminations on Chinua Achebe: The Art of Resistance*, coedited with Micere Githae Mugo for Africa World Press (2017). His current book project is coauthored with Quintard Taylor and examines the twentieth-century African American West. In addition, Ruffin has authored numerous articles, and has been an active consultant in regard to organizing curriculum, documentaries, public exhibits, and historical presentations on Africa and African Diaspora history and culture, including work with C-Span and the Smithsonian Institution and serving as U.S. Historian Delegate to South Africa.

PAUL J. P. SANDUL is Associate Professor of History at Stephen F. Austin State University, where he teaches courses in U.S. history, urban history, cultural memory, and public history. His publications include coediting and contributing to *Making Suburbia: New Histories of Everyday America* (University of Minnesota Press, 2015), which won an honorable mention for the Ray and Pat Browne Award for Best Edited Collection in Popular Culture and American Culture from the Popular Culture Association/American Culture Association (2016). His monograph *California Dreaming: Boosterism, Memory, and Rural Suburbs in the Golden State* was published by West Virginia University Press in 2014. His other publications include coauthoring two books for Arcadia Publishing about the California suburbs of Fair Oaks and Orangevale, as well as several book chapters and articles concerning suburbia, oral history, memory, race, and culture for such journals as *The Public Historian, Sound Historian, Agricultural History*, and *East Texas Historical Journal*.

M. SCOTT SOSEBEE is Associate Professor of History at Stephen F. Austin State University, where he teaches Texas, southern, and Latina/o history. He is also Executive Director of the East Texas Historical Association as well as Executive Editor of *East Texas Historical Journal.* He is the coeditor (with Kirk Bane and Charles Swanlund) of *A Lone Star Reader* (Kendall, 2015) and coauthor (with Michael Phillips and Keith Volanto) of a revised volume of *American Challenge: A New History of the United States* (Abigail Press, 2017).

INDEX

Made in the USA
Coppell, TX
05 January 2023

10460415R00152